Louis XIV's Botanical Engravings

This English edition first published in 2008 by

Lund Humphries
Gower House
Croft Road
Aldershot
Hampshire GU11 3HR

and

Lund Humphries
Suite 420
101 Cherry Street
Burlington
VT 05401-4405
USA

www.lundhumphries.com
Lund Humphries is part of Ashgate Publishing

A catalogue record for this book is available in the British Library

ISBN: 978-0-85331-990-0

Library of Congress Control Number: 2008921876

Produced by Editions de la Réunion des musées nationaux (RMN)

Publishing Director
Pierre Vallaud
Head of Book Publishing
Catherine Marquet
Commissioning Editors
Paule Foucret, Marie Lionnard
Editing of the French edition
Julie Houis
Graphic Design
Ursula Held
Production
Philippe Gournay
Repro
IGS, Angoulême

Printed by
Ingoprint, Barcelona

Translated by: Alayne Pullen on behalf of First Edition Translations Ltd
English edition edited and typeset by: Natalie Pomier
English edition proofread by: Elspeth Anderson

Louis XIV's Botanical Engravings

Alain Renaux

Ethnobotanist,
Centre for Functional and Evolutionary Ecology, CNRS, Montpellier

LUND HUMPHRIES

To Simone, Noémie, Alice and Jérémie, my four pillars of strength!
To Denis Dodart, a discovery and an inspiration.
To all those botanists and seafarers who had the courage to set sail and who paid for these perilous journeys with their lives.

Note to the reader

This volume, which contains fifty-eight of the three hundred and nineteen copper engravings preserved in the Chalcography Department of the Louvre, is inspired by Dodart's *Mémoires pour servir à l'Histoire des plantes*, published in 1676. It is by no means exhaustive but rather an invitation to continue on the path of discovery. In the past, man lived in closer contact with the natural flora of his immediate environment. He not only made practical use of what was around him – for food, medicine, clothing, and more – he also transposed into these elements of the natural world something of his own irrational, imaginative nature. And this has given us some wonderful stories ...

The plants have been identified from the engravings rather than from herbarium samples, hence the reservations regarding some of their names. At the period when Louis XIV's botanical engravings were being produced, certain botanical details were not necessarily determining criteria and it is likely that the scholars responsible for the work did not insist that the engravers pay special attention to these aspects.

The nomenclature used here is that of Michel Kerguélen. The order in which the plants are presented was decided by V.H. Heywood.

The Chalcography Department of the Louvre holds a national collection of more than thirteen thousand engraved plates. It was founded in 1797 with the bringing together of a number of copper engraving collections produced under the Ancien Régime, notably the Cabinet du Roi collection and that of the French Royal Academy of Painting and Sculpture.

The collections held in the Louvre Chalcography Department are administered by the graphic arts department of the Louvre Museum. Since its creation in 1895, the Réunion des musées nationaux (RMN) has been responsible for the commercial administration of the collections and produces and offers for sale prints taken from the original copper plates.

Acknowledgements

I should like to thank all those who have helped me either in identifying the plants or in re-reading my text: Cathy Bied, Max Debussche, Gilles Dutartre, Anne Gorgeon, Julie Houis, Daniel Jarry, Edouard Le Floc'h, François Munoz, Sandrine Pierrefeu, Elisabeth Pinard, Simone Renaux, Peter Schaeffer and John Thompson. Simone, François, John, Max and Edouard have been a great support to me and I should like to give them special thanks. Thank you also to Marie d'Hennezel for her warm support and to Béatrice de Boisseson, Marie Lionnard and Catherine Marquet for their faith in me.

Malus spectabilis (Aiton) Borkhausen (detail).
See p. 60

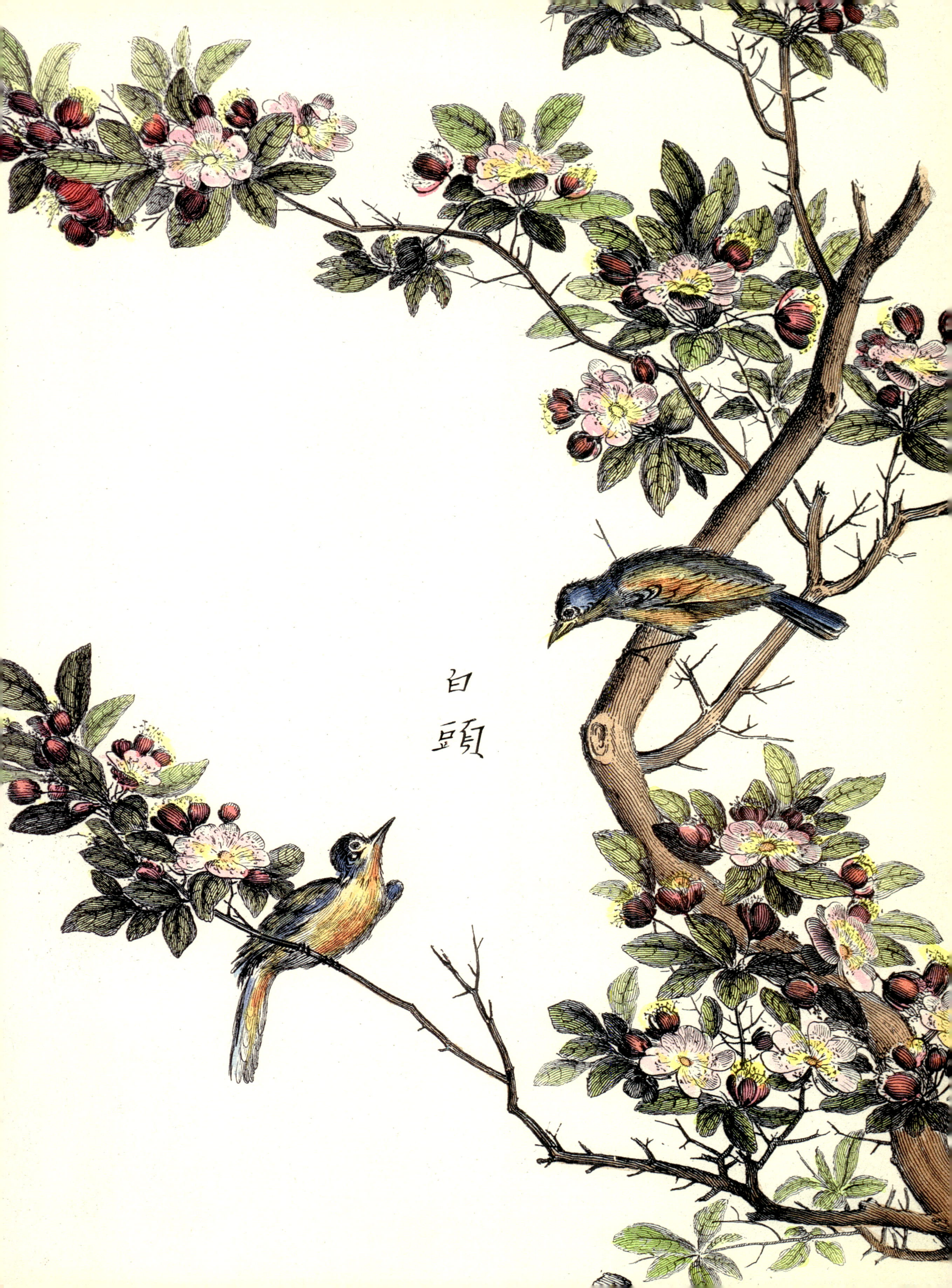

白
頭

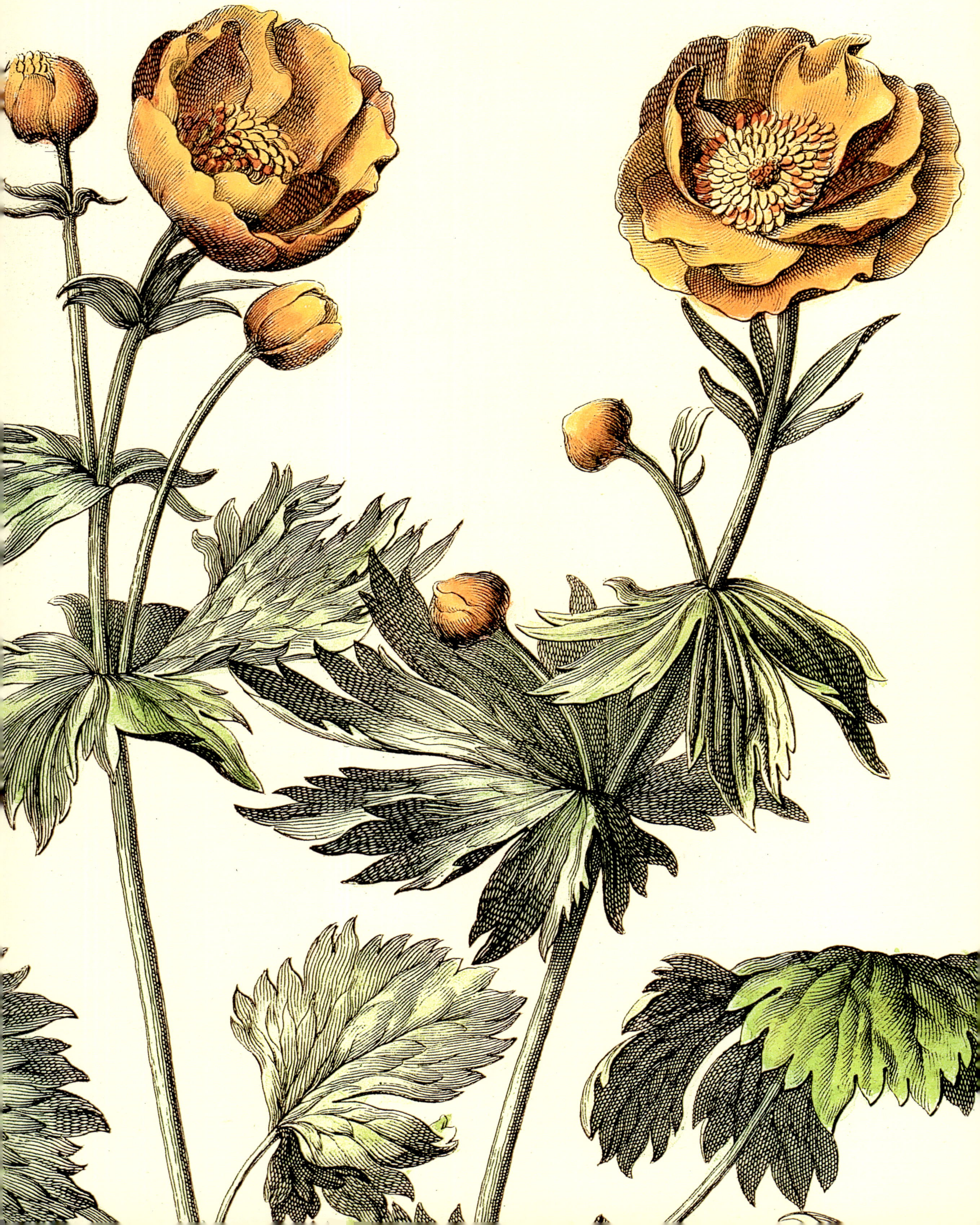

Foreword

The French Royal Academy of Sciences was founded by Louis XIV in 1666 at the instigation of his minister, Jean-Baptiste Colbert: 'Colbert selected a small group of scholars who gathered on 22 December 1666 in the King's Library, newly established in rue de Vivienne, and went on to hold twice-weekly work sessions'. This was the first meeting of the group, which comprised seven mathematician-surveyors, a qualified physician, six doctors of medicine (including two who were also physicians), four astronomers, two anatomists, a botanist, an engineer and another scholar whose discipline is unknown.

One of the projects undertaken by this brand new Academy was to compile a compendium of knowledge about plants in a volume entitled *Histoire des plantes*.

Claude Perrault, who was a physician to the king, an architect and one of the founder members of the Academy, entrusted this task to Denis Dodart, a botanist and also physician to the king. However, there was intellectual disagreement between the two men regarding the nature of living things.

Claude Perrault (1613–88), brother of Charles Perrault the storywriter, was famous in his time for his studies of anatomy, physics and natural history. He was one of the first to discover that sap circulated in both directions. He is also known for his designs for the east wing of the Palais du Louvre, the Paris Observatory and the Arc de Triomphe du Trône in the faubourg Saint-Antoine. He died from an infection contracted while dissecting a camel in the Paris Botanical Gardens.

Denis Dodart (1634–1707) became a doctor of medicine on 13 October 1660. Passionate about plants and chemistry, he practised his profession until 1666 when he was awarded the chair of professor of pharmacy at the faculty of sciences in Paris. Dodart was a prospective minister for foreign affairs but refused a number of prestigious positions in order to devote himself entirely to his passion. He was so intrigued by the capacity of plants to grow vertically that he wrote an essay on the subject. He became physician to the Duchesse de Longueville and to the widow of the Prince de Conti, and later medical consultant to Louis XIV. Dodart was accepted into the Royal Academy of Sciences in 1673.

A team of scholars

For this project, Dodart counted not only on Claude Perrault but on a number of eminent scholars in the fields of botany, medicine, pharmacy and chemistry: Samuel Cottereau du Clos, Claude Bourdelin, Abbé Jean Galois, Pierre Borel, Abbé Edme Mariotte, Nicolas Marchand and others.

Samuel Cottereau du Clos (1598–1685) was a physician to the king, a botanist and chemist, and took a particular interest in life sciences and the study of mineral waters.

Trollius europaeus L. (detail).
See p. 32

Claude Bourdelin (1621–99) was a pharmacist and chemist. He was one of the founder members of the Academy.

Abbé Jean Galois (1620–82) was originally a professor of mathematics, but left this post to teach Greek. He succeeded Denis de Sallo as editor of the *Journal des sçavans*, a publication founded in 1665 by Colbert which addressed issues associated with the sciences, religion and history.

Pierre Borel (1620–89), a prize-winning graduate of the Montpellier Faculty of Medicine, became a doctor of medicine at the age of twenty. A botanist and outstanding scholar, he was interested not only in botany but in chemistry, archaeology, optics, etc. Borel was a man with a passionate curiosity for all aspects of life. He entered the Academy of Sciences in 1674.

Abbé Edme Mariotte (1620–84) is better known for his works on the physics of fluids and gases (the Boyle-Mariotte law) than for those concerning anatomy and botany. He is known for his discovery of the blind spot in the eye. In his essay *De la végétation des plantes*, he sketches out an atomist theory, concluding that plants do not have souls and that the toxicity of plants depends not on the soil but on their own particular nature. Mariotte was also one of the founder members of the Academy.

Nicolas Marchand (?–1678) was a doctor of medicine of the University of Padua, chief botanist to Gaston de France (Gaston d'Orléans), director of plant cultivation at the Jardin du Roi and one of the founders of the Academy. His son Jean followed in the footsteps of his botanist father, taking over his seat in the Royal Academy when it became vacant.

It was decided that, to avoid any risk of confusion, the *Histoire des plantes* should contain a detailed description of each plant along with an accurate illustration. A simple herbarium alone would not be sufficient for an illustrator who was not a botanist to represent plants at their various stages of development, or even in their original colours. Marchand therefore took on the task of assembling a collection of seeds, sowing them in the Jardin du Roi and leaving them to grow. As no fewer than three hundred and nineteen species were selected, this was an extremely lengthy undertaking.

Master engravers

Next, a group of highly skilled engravers, masters in the art of burin engraving and etching, had to be found.

Abraham Bosse (1604–76) was one of the best. He had learned the basics of engraving under Melchior Tavernier, a French book and map engraver working in Paris and master of the art of copperplate engraving. Bosse met two accomplished etchers from Lorraine, Jacques Bellange (much of whose work has been lost) and Jacques Callot who was extremely influential in the development of engraving. Callot is known for his war scenes, in particular one showing a tree with hanged men entitled *Les Misères de la guerre* which used to be reproduced in school textbooks.

The training Bosse received from these two masters was fundamental to his development as an engraver and they were a strong inspiration on his work. However, his acquaintance with the architect Girard Desargues was even more significant. Desargues introduced and taught the concept of linear perspective with the line receding to vanishing point. This was a revelation for Bosse who attended Desargues' classes and became one of the most ardent champions of this method. He even published essays on the subject.

When, in 1648 under the regency of Anne of Austria, Cardinal Mazarin founded the Royal Academy of Painting and Sculpture, Bosse became a founder member, along with such celebrated artists as Charles Le Brun and Philippe de Champaigne. The aim of this Academy was not so much to create an elite body that would impose artistic principles but to give status to artists, raising the standing of its members. Academy members wanted to be seen as distinct from craftsmen artists, painters, sculptors and printers, who were members of the all-powerful guilds and brotherhoods of St Luke, whose power extended throughout western Europe.

Colbert, attracted by the idea of placing this group of artists at the service of the state, took an interest in the Academy and this won it the recognition it needed. Colbert became its patron and appointed Le Brun as director. However, dissension soon arose between Bosse and Le Brun over the famous rules of perspective and the direction art should follow, ending in 1661 with Bosse being excluded from the royal institution.

Bosse had discovered a liking for naturalist observation and the technical nature of graphic art. Guy de la Brosse, founder and director of the Jardin du Roi, asked him to engrave a thousand plates of plant drawings for his work on medicinal plants, the *Catalogue des plantes cultivées à présent dans le Jardin Royal des plantes médicinales*. Each plate was to show the general features of the plant – the flower, fruit and any unusual details. The death of La Brosse in 1641 put an end to the project but Bosse had already built himself a solid reputation as an artist-engraver. It was therefore natural that he should be invited to join Dodart's great project, for which he produced forty-seven plates, though only ten of them were published.

Little is known about Louis de Chastillon (1639–1734). He contributed some thirty engravings to the collection, the best known of which seem to be the two 'Chinese hawthorns' which are in fact an apple tree and a plum tree.

Nicolas Robert (1610–85), another important figure of the period, was an engraver and miniaturist. The quality and refinement of his embroidery designs first brought him to the attention of Gaston d'Orléans, brother of Louis XIII. He excelled in the art of painting flowers, birds and other animals. Robert produced the twenty-nine flowers of the *Guirlande de Julie*, a collection of madrigals which the Duc de Montausier wrote or commissioned from certain poets as an expression of his love for Julie-Lucine d'Angennes de Rambouillet. A protégé of Gaston d'Orléans and Colbert, Robert was a contributor to the collection known as the *Vélins du Roi*, producing four hundred and seventy-five of what is now a six thousand strong collection.

Of the three hundred and nineteen plant engravings chosen by Dodart, at least half are by Robert. All are burin engravings and etchings.

From copperplate engraving ...

It appears that copperplate engraving and along with it the printmaking technique of engraving were invented by the Italian Maso Finiguerra in 1452. This technique uses only burin or drypoint on a relatively soft metal sheet known as a pattern. The pattern is often made of copper or zinc, sometimes of iron. (The results achieved are so fine that the technique is still used today in the manufacture of dies for the printing of postage stamps and banknotes.)

Pressing the point down into the metal with varying intensity, the engraver produces a reverse image. The shading comes from the density of the line. The engraver then checks his work by printing it as he goes along, correcting it and improving it until he is satisfied with the result. When using iron or soft steel, the metal has to be hardened by tempering, in the same way as tempered steel.

To produce a print, the pattern is coated with ink and the surplus removed with a special cloth or absorbent paper. The engraver then finishes applying it with the palm of his hand so that it penetrates right down into the hollows, which are known as cuts. A sheet of damp paper is then placed on top and pressed against it with the appropriate tool.

... to copper etching

This technique calls upon the mordant properties of nitric acid on metal. The process may be simple but there are different methods and the art of copper etching is a difficult one. A plate of copper, bronze or another metal is coated with a hard varnish and this is then cut into with a metal point to produce the desired image. The prepared plate is then plunged into a bath of nitric acid diluted with water for a set period of time. The acid eats into the metal where it has been cut and this produces the engraving.

Copper etching dates back to the Middle Ages. Arab gold and silversmiths used it in jewellery manufacture, to decorate weaponry and in copper ornamentation. In the fifteenth century, the technique was applied to printing. Albrecht Dürer (1471–1528) made woodcuts for his *Small Passion* and *Great Passion* but produced another just as impressive copper etching of the *Passion*.

The technique developed extensively under Jacques Callot who introduced a burin with a triangular tip, making it possible to produce both full and fine lines, in much the same way as the quills once used by schoolchildren. Callot discovered the hard varnish used by lute makers, which was a great help in the engraving process itself as it allowed the engraver to place his hands on the work – an impossibility with soft varnish. He also perfected the multiple bath technique in which certain areas are dipped and then protected with varnish before the plate is re-dipped, the process being repeated until the desired effect is achieved. This allows variation in the depth of the cuts, the contrast and highlights.

Dodart was determined that the illustrations for the *Histoire des plantes* should be produced as etchings: 'We prefer Etching to any other form of engraving because it allows greater freedom, is easier and more rapid and is just as sharp as copperplate providing it is well executed.'

With the admission of Bosse to the Royal Academy of Painting and Sculpture, engraving became a major art form, on a level with the two other arts, as confirmed by the Edict of Saint-Jean-de-Luz of 1660. It continued to develop over the centuries thanks to the brilliance of the engravers and the many painters who chose to express themselves in this form, from Rembrandt to Watteau and Degas and, more recently, Matisse, Picasso and others.

Anchusa cretica Miller (detail).
See p. 92

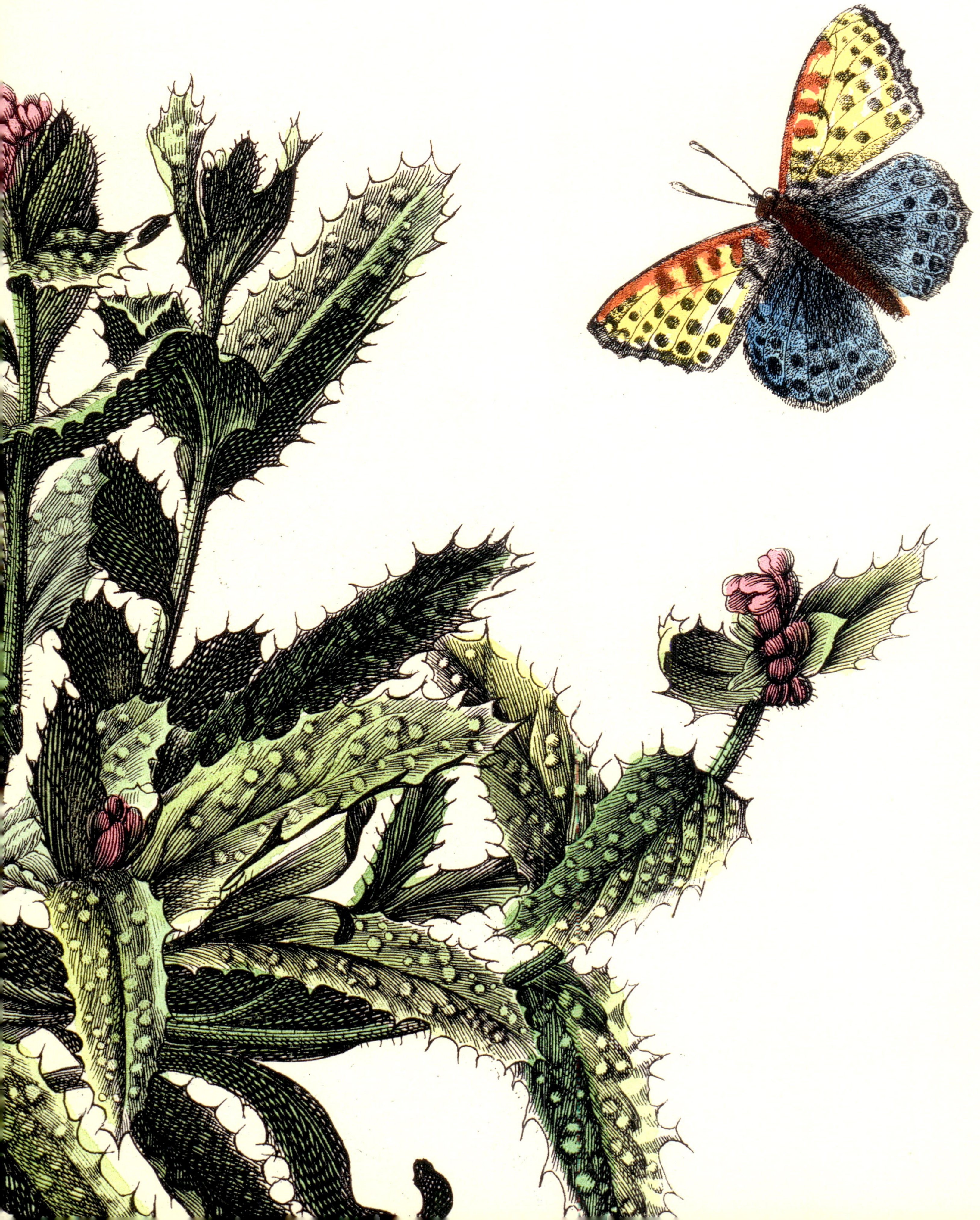

Plants in everyday life

Unlike today, in the seventeenth century plants played an important role in many aspects of everyday life. In fact, ethnobotanical studies of the period between the First and Second World Wars reveal the importance they still had at that time. In the Cévennes region, and in most mountainous areas of France, people still collected pellitory, nettle, horsetail and other plants for use in washing up, wood ash, soapwort and ivy for washing clothes, willow, hazel, oak and viburnum for basketry, and various timber species such as chestnut, walnut, oak, ash, birch and beech for buildings, furniture, carts, boats, barrels, clogs etc. Plants were also used to provide light, for human and veterinary medicines and for dyeing textiles.

By the start of the seventeenth century, medicine appears to have developed very little since the Middle Ages, despite the anatomical discoveries of the Renaissance. There were however true 'geniuses', such as the doctors Girolamo Frascatoro, Paracelsus, Santorio and the surgeon Ambroise Paré. The Spanish doctor Miguel Servet discovered that venous blood is purified by the lungs before returning to the heart. (Yet, because of his opposition to religious dogma, Calvin had him burned alive, with his books, in Geneva.)

The study of medicine took three years, each year being divided into two semesters known as the 'grand ordinaire' and the 'petit ordinaire'. At the end of their studies, students were examined on anatomy, diagnosis and therapy and had to answer questions on the work of Hippocrates and Galen. Teaching on disease was based on the four main humours produced by the liver: blood, pituite or phlegm, atrabile or black bile – also known as melancholy, from the Greek *melanos* (black) and *chole* (bile) – and yellow bile or choler. This was the era of blood letting, purges and enemas, and the mortality rate remained very high.

Little by little, this century witnessed some important discoveries. These included: the improvement of the Jansen microscope by Antonie van Leeuwenhoek, who described red blood cells; the elaboration of the concept of the cell by Robert Hooke; the discovery of the ovarian follicle by Reinier de Graaf; of spermatazoa by Louis de Hamm; and of blood circulation by William Harvey. Incredible though it may seem, certain eminent French doctors – Jean Riolan and Guy Patin (dean of the faculty of medicine in Paris) – in fact challenged Harvey's discovery.

Plants were the chief material used in medicine and therefore played a crucial role. They were used mainly *in totum* (that is to say in their entirety), following the precept of doctor Ibn Sina, known in Europe as Avicenna (980–1037 AD), according to whom 'the whole is greater than the sum of the parts'. This knowledge was founded on ancient tradition, passed on meticulously from generation to generation, and just as carefully transcribed in fundamental works, such as that of Jacques Daléchamps (1615) and many others.

The ideas of Paracelsus – that patients should be treated according to their particular ailment and that the active principle should be isolated – were beginning to gain ground. And his celebrated dictum remains true today: 'Every remedy is a poison, with no exceptions. It is all a matter of dosage.'

Anemone nemorosa L. (detail).
See p. 30

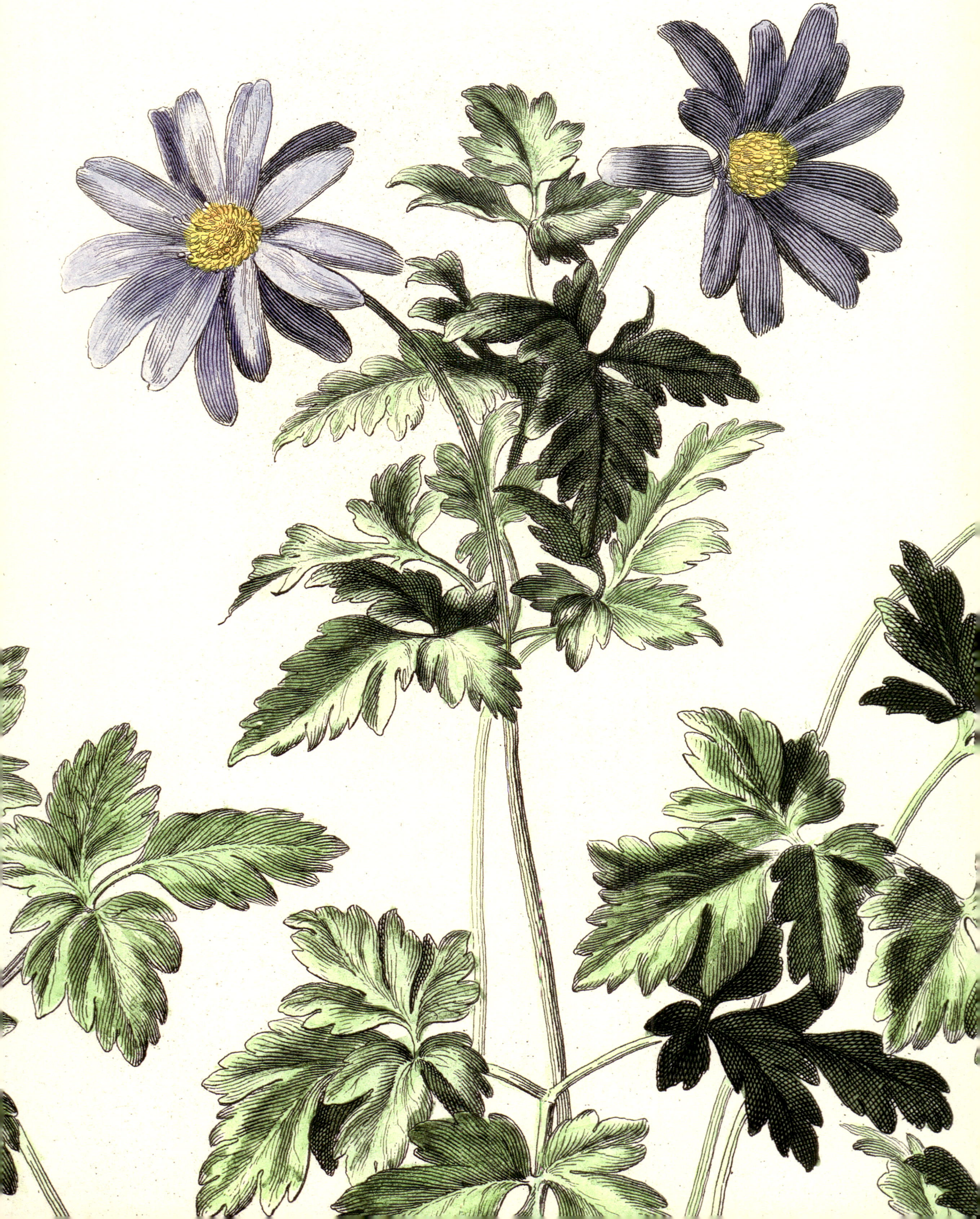

The first exotic plants

In circumstances such as these it is easy to understand why people were so receptive to the introduction of new plants arriving in Europe from the 'West Indies' and from other parts of the globe. From the sixteenth century onwards, missionaries like the Spaniard Francisco Lopez de Gomara (chaplain to Hernán Cortés), the Italian merchant Girolamo Benzoni, the French priest André Thévet (later chaplain to Catherine de Medici) and others, published works on the 'West Indies' which included descriptions of plants and their uses. They also brought back plant specimens from their voyages. In fact, the history of botany is rich in strange, racy and fascinating stories.

Thévet, for example, returned with the seeds and plants of a very attractive and decorative member of the family *Solanaceae*, which was in fact tobacco. It was known as *pétun* (the native word for tobacco and from which the name 'petunia' derives). Thévet grew the plant on his return and distributed it to his friends. When Christopher Columbus landed in the Bahamas, the Arawak Indians made him a gift of the leaves of an unfamiliar plant. Not knowing what to do with these he ended up throwing them away. However, a month later, his lieutenants Jerez and Torrez, on an expedition to the island of Cuba, saw native people in a village burning the leaves of this plant and inhaling the smoke with the aid of a hollow stick called *tobago* or *tobaca*. Intrigued, they experimented with this themselves and thus probably became the first European smokers in history. Unfortunately for Jerez, when he returned to Spain, religious inquisitors saw the smoke issuing from his mouth, accused him of making a pact with the devil and imprisoned him for seven years. He must have felt himself fortunate to escape with such a light sentence for a charge of this nature, as the clergy at this period were far from enlightened. As the practise of smoking became an increasingly common social habit, he was eventually released. (In nineteenth-century China tobacco smokers were beheaded!)

Much to Thévet's displeasure, tobacco became popular in France not through him but through Jean Nicot de Villemain, French ambassador to Lisbon at the time. Villemain received a small supply of leaves which he used at court to ease the migraines suffered by Catherine de Medici.

The arrival of new plants and the interest they aroused throughout much of Europe led to the creation of botanical gardens: in Padua in 1545; in Pisa in 1546; Leiden in 1577; Montpellier in 1593; Paris in 1597 or 1591 (depending on whether one reads the final figure as a 1 or a 7 and according to the assertions of Olivier de Serres) ... Be that as it may, the botanical gardens of Montpellier have always been considered the earliest royal gardens still surviving today. Created by an edict signed by Henry IV of France at Vernon on 8 December 1593, they served as a model for later French botanical gardens: 'Henry by the grace of God, King of France and Navarre, to our loyal and beloved counsellors ... that the said Belleval is required to obtain and create a garden in which are to be put simples and all manner of plants that may be obtained, both foreign and domestic ....'

An area had to be set aside on ships for the plants and sufficient water brought on board to supply them on the long journey that awaited them. Columbus's ships, the *Santa Maria*, *Niña* and *Pinta*, took nearly two and a half months – from 3 August to 12 October 1492 – to reach the island of San Salvador. The expedition that Thévet joined in 1555 took six months to reach Brazil. Arrangements also had to be made at the ports to store these plants for a month or two. This was the start of the 'storage gardens' at Lorient, Brest and other ports. These usually consisted of heated glasshouses of the orangery type.

Arisaema triphyllum (L.) Schott (detail).
See p. 128

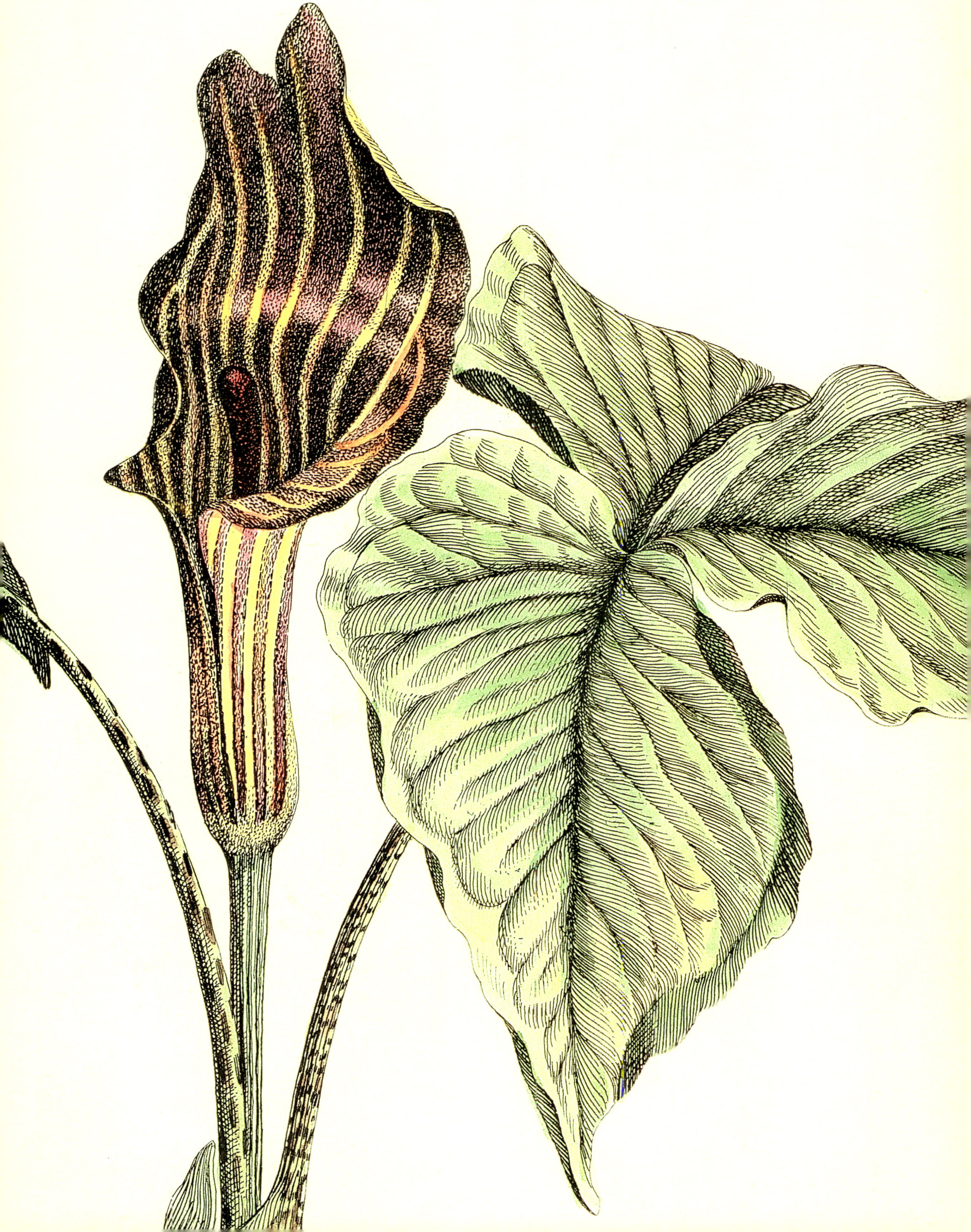

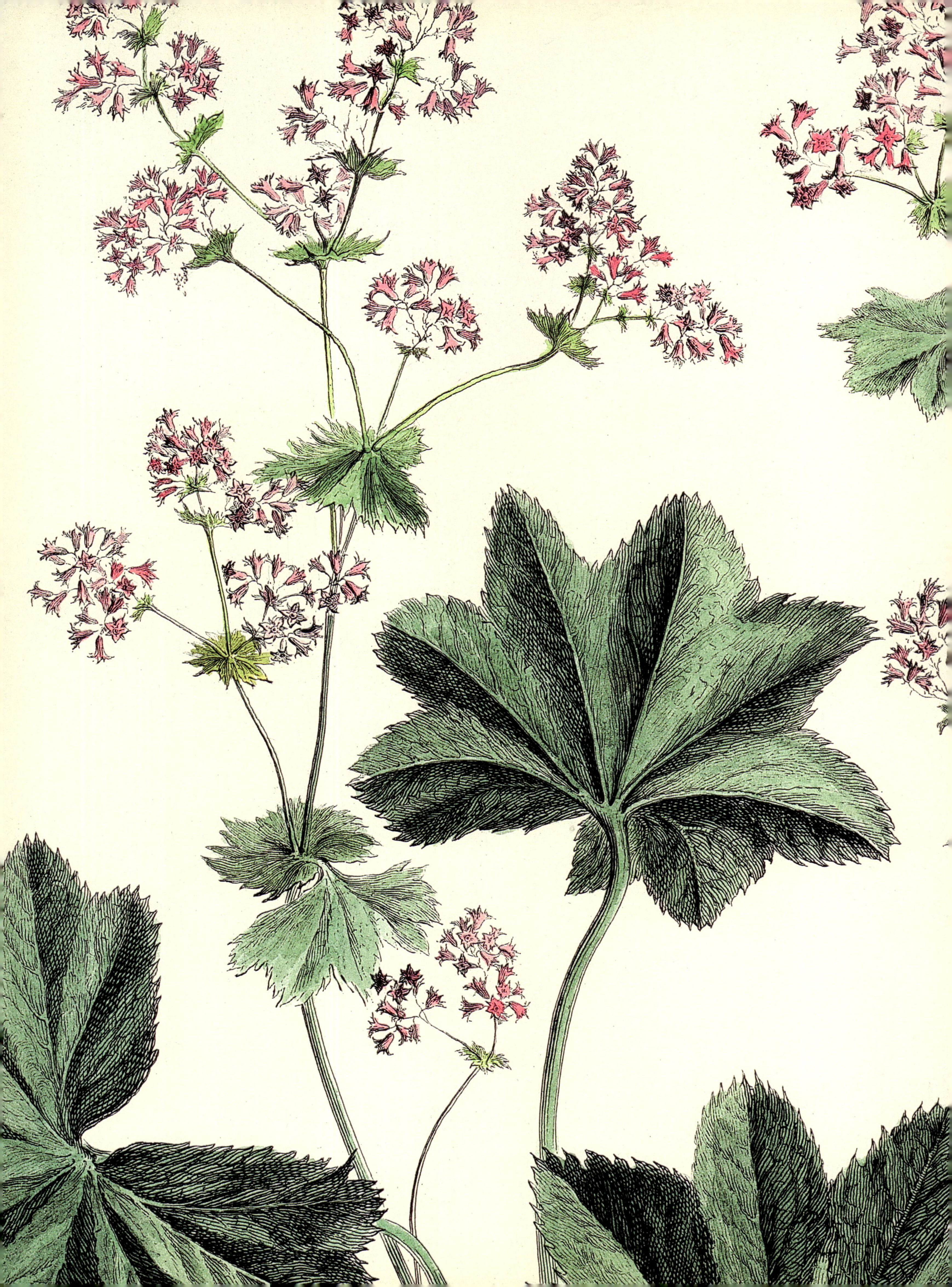

A great healer

According to Peruvian legend, in ancient times there existed a great lake near the village of Loxa. Following a major earthquake, the trees that grew along its shore fell into the water. A local native suffering from fever drank this water and was cured.

In 1630, don Francisco López de Cañizares, the *corregidor* of Loxa, fell ill with a fever and a local native treated him with the bark of a tree known in the Quechua language as *kinakina* ('cinchona', from which the name 'quinine' is derived). In 1633, Antonio de la Calancha (known as 'Father Calancha'), an Augustinian monk on mission in Peru, refers to 'a tree known as "the fever tree"... in the Loxa area, the cinnamon-coloured bark of which, when ground to a powder in weight equivalent to two small silver coins and made into a drink, is used to treat fevers'. Many years later, Joseph de Jussieu, who visited Loxa in 1739, reported that a Jesuit priest was also cured by this plant.

In 1638, Countess Ana de Chinchón, wife of the viceroy of Peru, fell ill. Don Francisco López sent his doctor, Juan de Vega, to attend her with the precious powdered bark. When she recovered, the countess distributed the bark generously among her acquaintance and it became well known. From that time onwards it was called 'the countess's powder'. On her return to Spain the countess died, but her husband, don Luis Jerónimo Fernández de Cabrera y Bobadilla, Count of Chinchón, continued to promote the plant.

At roughly the same time, in 1640, the Jesuit general Mutio Vitelleschi had a large quantity of this bark sent from Peru and tested its efficacy in many European cities, after which the remedy also became known as 'Jesuits' powder'. From Spain to Italy, Cardinal Johannes de Lugo passed the remedy on to his colleague Mazarin to administer to the young Louis XIV who was ill at the time. Years later, having contracted a serious fever during the war, Louis XIV was once again treated with quinine by his doctor, Antoine Vallot. The bark became increasingly used, despite opposition from the short-sighted but influential Guy Patin.

An English apothecary's apprentice by the name of Talbor was shrewd enough to see in it an opportunity to become rich and famous. He devised a quinine-based preparation and concealed the bitter taste with other plants. He was successful in treating Charles II, curing him of quartan fever (malaria) and thus became his physician. Talbor also went to France and cured in turn the Dauphin, Colbert and other distinguished French figures. At the time, marshy areas were widespread throughout Europe, and in France in particular – in fact Versailles was one gigantic marshland before it was drained. And malaria took a heavy toll. (The air was bad – hence *mala aria* – and people caught fever from the marshes – *palus, paludis* in Latin – hence paludism.)

Louis XIV offered Talbor the title of 'chevalier', together with two thousand gold louis and a commensurate allowance to reveal his secret, which he eventually published in 1682. In fact the response of fever to 'Jesuits' powder' became the earliest diagnostic test and in 1712, Dr Francesco Torti used it to distinguish paludal fever from other fevers. In 1742, the celebrated Swedish botanist Carolus Linnaeus named the tree Cinchona, in memory of the countess. The genus contains many species of interest to man. The principal alkaloid, quinine, was isolated in 1820 by Pierre Pelletier and Joseph Caventou.

Alchemilla vulgaris L. (detail).
See p. 52

At the time of Dodart, many of the plants familiar to us today were already here, either in their familiar form – the love apple or tomato, the pepper, *papas* or Clusius's potato (even though it was not yet eaten), sunflowers etc. – or in the form of manufactured products – cane sugar, tea, coffee etc. Many of the plants were introduced into Europe by the great travelling botanists such as Charles Plumier (1646–1706), Joseph Donat Surian (1650–91), Michel Sarrazin (1659–1734), Auguste Lippi (1678–1705), Louis Feuillée (1660–1732), Amédée-François Frézier (1682–1773) and others. Frézier, as if predestined by name – Frézier sounds like *fraisier* (strawberry plant) in French –, returned from Chile with a small plant with reddish-white 'fruit': this was the Chilean white strawberry, *Fragaria chiloensis* (L.) Miller. A young Antoine Nicolas Duchesne (1747–1827), one of the royal gardeners, asked Frézier's permission to hybridise it with a strawberry plant from Virginia, *Fragaria virginiaca* (Miller) Staudt, and produced the famous Plougastel strawberry.

However, many scholars and sailors never returned from these perilous voyages. Almost a century later, on expeditions with Captain Cook, Captain Clerke died of tuberculosis and the botanist Spörring and the astronomer Green died from amoebic dysentery. Cook himself was killed in 1779 by native Hawaiians. The ships that set sail on these expeditions had names such as *Resolution*, *Endeavour* and *Discovery* – and the chances of returning safely from their voyages were a thousand times worse than those of returning safely from a flight on board any of the spaceships later named after them.

From first steps in botany ...

One hundred and seventy-four years had passed since the official discovery of America and many plants had been brought back, placed in herbariums or planted in botanical gardens, royal gardens and the gardens of the nobility. Modern botany was still in its infancy and although Latin names were beginning to be shortened, some to the binomial form, the determining criteria were far from reliable.

From the early Greek botanists, such as Aristotle and Theophrastus, to the Renaissance, there had been very little progress in this respect. In fact, more than three centuries before Christ, Theophrastus had, through his observations and methods, left considerable data that continued to be used by the botanists of the fifteenth century and as late as the nineteenth century. This is true of the distinction between monocotyledons and dicotyledons. But, most importantly, there was a major philosophical point at issue: Theophrastus had disseminated the idea that plants, and nature as a whole, acted in accordance with their own plans and not in order to be useful to humans. This concept, though obvious to us today, was revolutionary at the time. In Arab, Indian and Chinese civilisations, plants were identified according to their use as food or medicine and nothing more. The Middle Ages saw some slight progress from classifying plants according to where they grew and their uses to classifying them as trees or herbs.

... to the birth of taxonomy

From the sixteenth century onwards, the exploration of new territories began to have a considerable impact on botany. It is easy to imagine the fascination felt by scientists, clergy, noblemen, and probably ordinary mortals too, for these unfamiliar plants with their important medicinal properties.

It is rather as if, today, strange plants capable of curing diseases believed to be incurable were brought back to earth from some distant galaxy. This was the century that saw the flowering of works

Aesculus hippocastanum L. (detail).
See p. 78

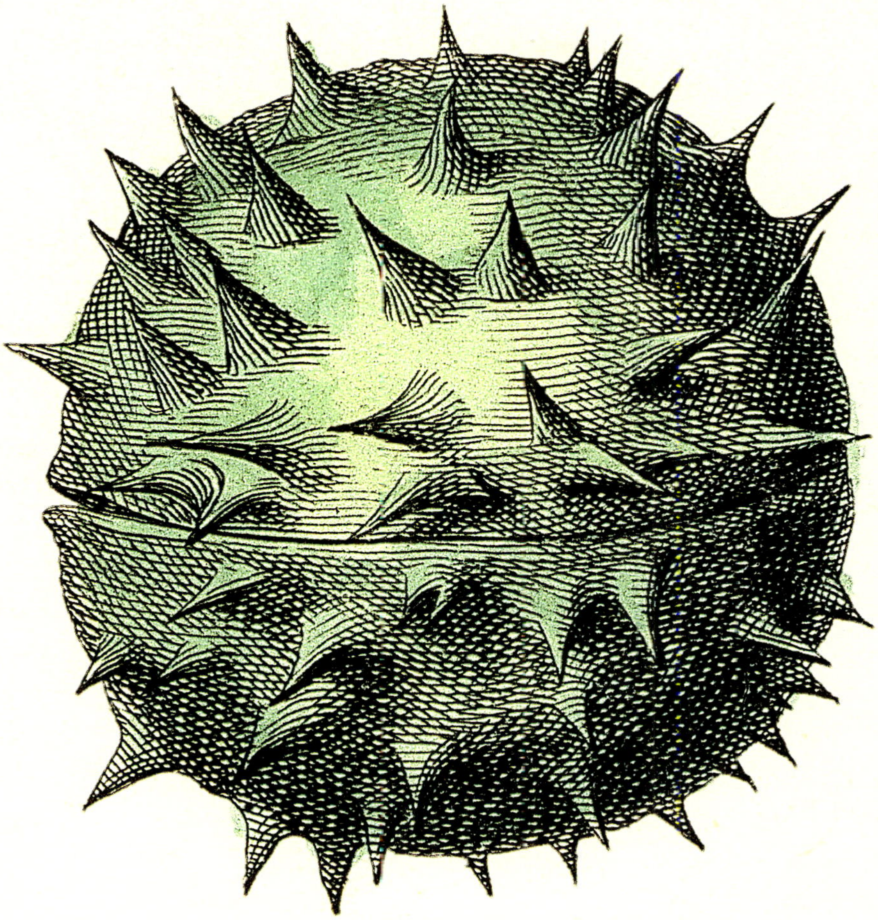

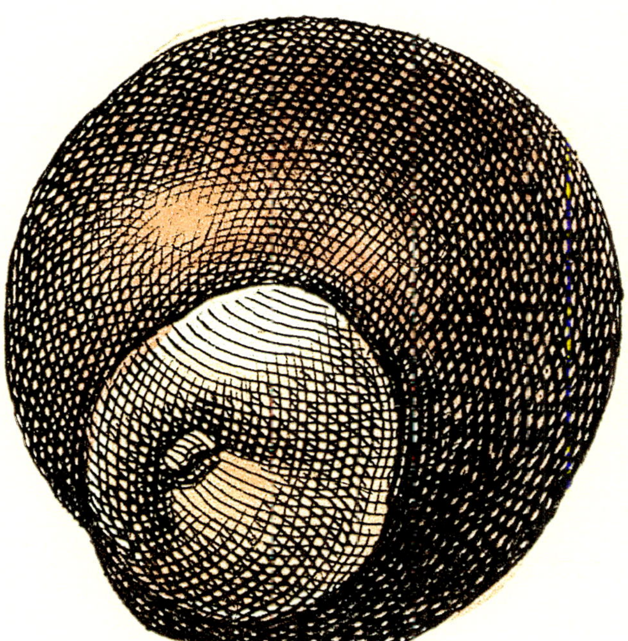

by the forerunners of modern botany such as Fuchs, Mattioli, Bauhin, Daléchamps, Bock. Slowly and gradually each contributed his building block to the edifice of botany, a block that to some may have seemed small but which often proved to be a key link in the evolution of knowledge.

In the seventeenth century, flowers began to be seen as an important element in classification. Some of the most significant botanists of the period include:

Joachim Jung (1587–1657), who was an unparalleled observer of plants. His works proved to be a major influence on naturalists of his period and those who followed later. He introduced precision in observation and in the terminology of the different parts of a plant. Jung introduced new terms including petiole and peduncle, digitated (palmate) and pennate, alternate and opposed leaves and perianthus for the coloured part of the flower. He distinguished between single and composite flowers, used the words *stamina* for stamen, and stylus and stigma, although plant reproduction remained a mystery at that period. He even identified the presence of an embryo at the heart of the seed, the *cor seminis*, and its inferior or superior position within the seed. His book *Isagogue phytoscopia* was published a year after his death.

Rivin (1652–1723), who, in his *Introductio ad rem herbarium* of 1690, put forward a classification system based on the form of the corolla.

John Ray (1627–1705), who contributed to the classification of plants into monocotyledons or dicotyledons by identifying the number of cotyledons in the seed, and later took a particular interest in flowers and fruit.

Pierre Magnol (1638–1715), who made a fundamental contribution by introducing the term 'family' and by considering plants in their entirety, including the roots, stems, leaves, flowers and seeds. Later, Michel Adanson (1727–1806) returned to this fundamental concept and rejected the classification systems developed by Linnaeus, Tournefort and Ray paving the way for the phylogenetic classifications of today.

Joseph Pitton de Tournefort (1656–1708), who continued with the definition of corollas while introducing clear and fundamental concepts of genus and species. In 1694 he published his three-volume *Eléments de botanique ou Méthode pour connaître les plantes* in French and then a five-volume Latin edition entitled *Institutiones rei herbariae*. His system includes flowers that are petalled and apetalous, single and composite, regular and irregular as well as various plant families – such as papilonaceae, liliaceae, labiaceae, rosaceae and umbelliferae – whose names persist today in the classic floras of France, while others such as infudibuliform, personate, anomal and flosculous have disappeared.

Unfortunately his research came to an abrupt halt when he was fatally wounded in a carriage accident (in the street that bears his name in Paris's 5th district). However, this great botanist's work had paved the way for the advent of a man considered to be the father of the nomenclature of living things: the Swedish naturalist and physician Carolus Linnaeus (1707–78).

A collective endeavour

Dodart and his team found themselves faced with the immense task of organising all this information while humbly admitting the inadequacies of the systems applied. They were only too aware of the scale of the task before them, of the rigour that must be exerted to complete it successfully and of the need to work as a team. Although the abilities and preferences of each scholar may have been limited to a particular field, all shared a sense of unity that went beyond their sometimes divergent points of view. The end result, a written account aimed at a more or less scholarly audience, had a duty to reflect the coherence of these endeavours.

Each element received precise and meticulous attention – from the gathering of general information to the minutest detail of an engraving. Those scholars who were most expert in botany became a constant presence in the engravers' studios.

A first volume appeared in 1676 under the title *Mémoires pour servir à l'histoire des plantes, Dressez par M. Dodart, de l'Académie royale des Sciences, Docteur en Medecine de la Faculté de Paris.*

The united spirit of this team endeavour is clearly reflected in the note to the reader: 'This book is the work of the entire Academy. There is none among those of whom it is composed who has not been its judge and who has not contributed at least some opinions. Messieurs du Clos, Borel, Perrault, Galois and Mariotte have examined it most particularly and the content of this work is the result of the proposals, experiences and reflections of several members of this Assembly ... It is to the attentions and endeavours of Monsieur Marchand that we owe almost all the rare plants presented here and it is he who has given us the names of Plants not yet described, their Descriptions and their Cultivation.'

As well as reading all studies ancient and modern on the subject, the project also involved the cultivation of plants so that they could be better described, 'and according to their parts, at their different ages and at different seasons ... We shall therefore continue this work this spring on young plants of which we do not yet have sufficient experience ... By doing this we hope to compile material to give to the public year by year. Anyone who has fully understood the scale of this work and all the endeavours that must be undertaken to fulfil it (with which we shall not bother the public), will easily judge it to be a great achievement if we are able to do what we propose here ...'.

However, *Mémoires pour servir à l'histoire des plantes* remained the only volume published. What then occurred between the original, major project and this partial publication? There do not appear to have been any obstacles of a political or financial nature. Was is the sheer enormity of the project that prevented its completion? Or were there insurmountable internal conflicts?

In 1694, that is to say eighteen years after the appearance of this first and only volume, the Academy's project was abandoned. Jean Marchand, who succeeded as director of the Jardin du roi when his father died in 1678, lost his post that same year. Despite everything, he decided to continue the work begun and located the plate engravings but died before he was able to complete the project.

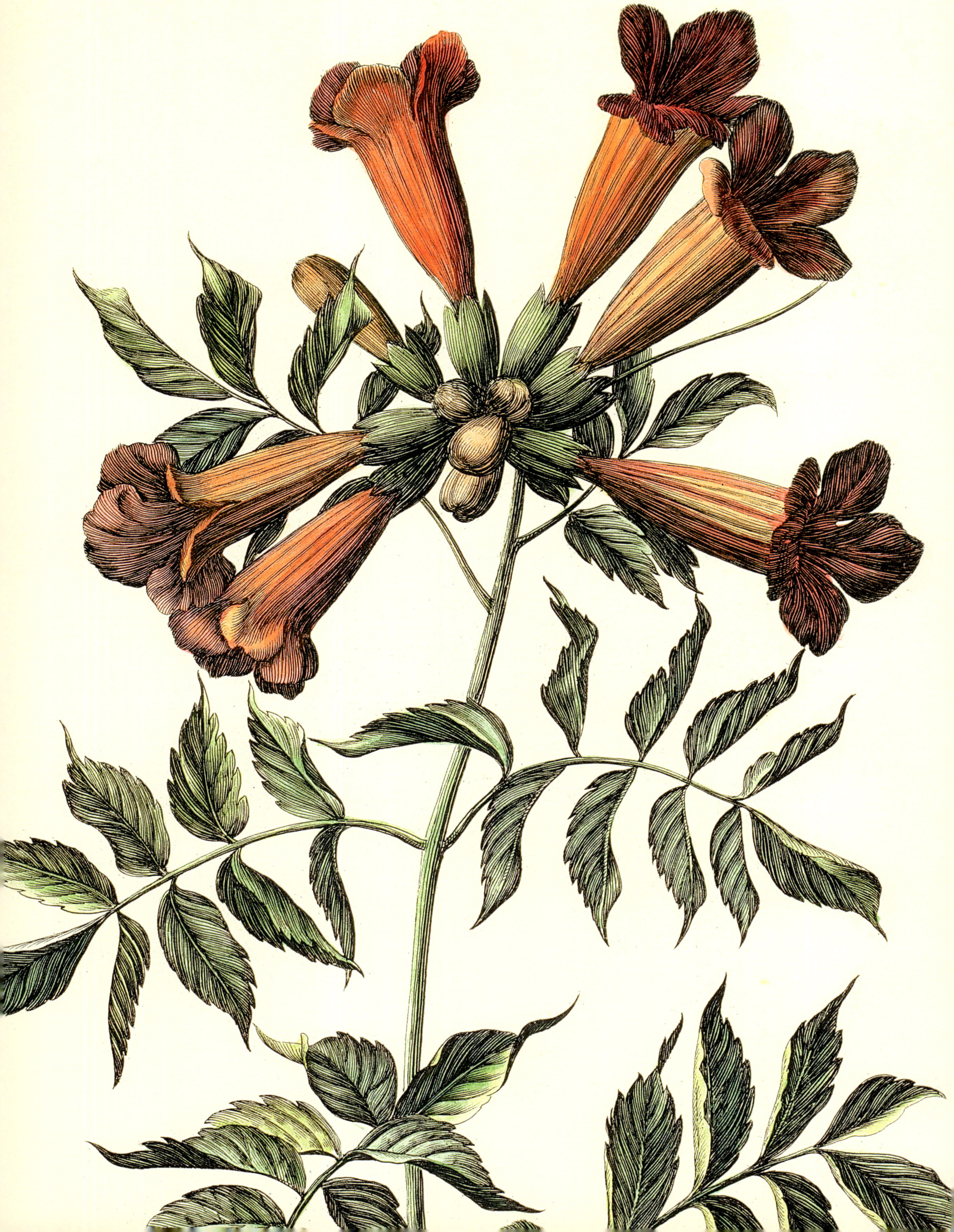

Plates

Campsis radicans (L.) Seem. (detail).
See p. 122

Aconitum napellus L.

RANUNCULACEAE

Handsome and majestic, this aconite – also known as monkshood, wolfsbane, Venus's chariot and friar's cap – is found in mountainous areas at an altitude of between 500 and 2,800 metres (1,640–9,000 ft), in cool, damp situations, often close to sheepfolds. Depending on the region and the location, the subspecies *vulgare*, *corsicum*, *lusitanicum* (from Portugal but found in France) and *napellus* may be found. The latter includes two natural varieties: *napellus* and *giganteum*. It may seem excessive to the layman that one plant name should contain the term *napellus* three times, but in botany when different varieties are being described, the taxon will include the same name three times.

The exact etymological origin of the term 'aconite' is unknown, despite the various interpretations given by the Ancients, but it seems the Greek word *akoniton* was applied to a number of poisonous plants while *napellus* comes from the Latin *napus* meaning turnip, related to *rapum* (rape).

Aconitum napellus L. contains a formidable poison. This property of the plant was well known to our remote ancestors who, with a clear concern for efficiency, coated the tips of their arrows and lances with its juice and with that of the yellow wolfsbane aconite (*Aconitum lycoctonum* L. and its two subspecies, *lycoctonum* and *neapolitum*). The other aconites found in France – *A. anthora* L., *A. burnatii* Gayer, *A. variegatum* (subsp. *paniculatum*, *pyrenaicum*, *valesiacum* and *variegatum*) – are just as formidable.

The plant is certainly extremely toxic but in low doses known only to pharmacists it can be used medicinally. In fact its medicinal properties are much valued in the treatment of cardiopulmonary disorders, spasmodic coughs, and facial neuralgia.

The essence of the drug – aconitine in particular, which is a very powerful diterpenoid alkaloid – is concentrated in the young, tuberous roots. Its anaesthetic action first paralyses the sensory nerves; it will eventually stop the heart and lungs, producing the impressive effects described in toxicology manuals.

Daléchamps (1615) gives a colourful description of it: 'It is a poison that kills those who drink it, the largest dose being half a dram [author's note: 3.24 grams], and even less than this can, in my opinion, cause death. The counter-poison is Mouse of Napellus: a mouse fed on this plant ... such an animal, known as *Mus napelli*, is resistant to *Napellus* and can withstand its poison when eaten. *Napellus* is one of the most pernicious poisons in existence. It causes the lips of those who drink it to swell and the tongue also, making their eyes start from their head, and causing their head to spin and their heart to pound ... To such an extent that, as Matthiol says, it will kill a person who holds the root in his hand while heating it. He even says that there are shepherds who have died after using *Napellus* root in place of a skewer to roast small birds.'

In the mountains, thin willow branches can be used as skewers, providing they are not mistaken for laburnum which, though highly ornamental, is extremely poisonous!

Aconitum napellus L. Engraving: Nicolas Robert
National Museum of Natural History, Paris.

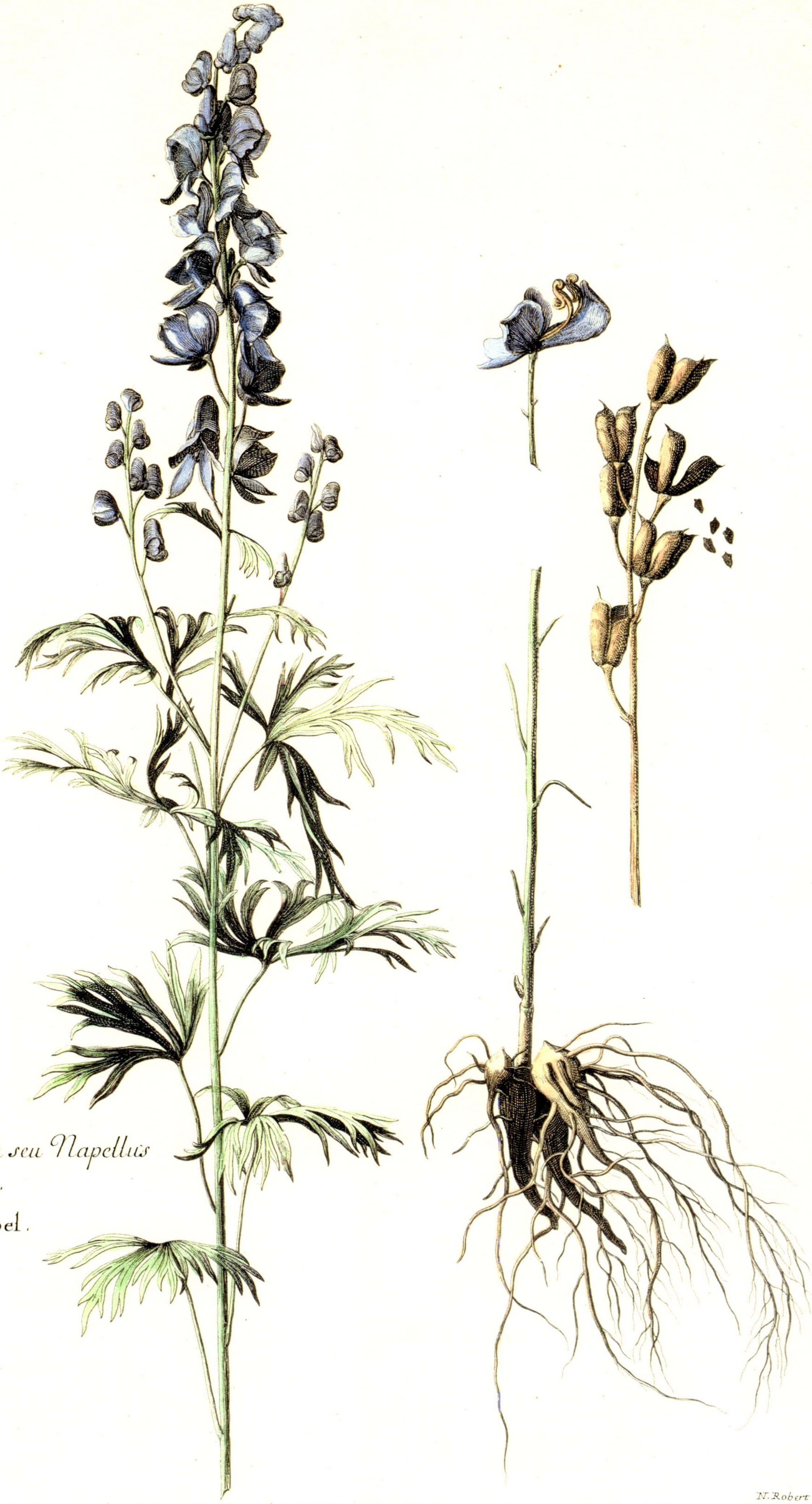

Aconitum coeruleum seu Napellus vulgaris.

Aconit ou Napel.

N. Robert del. et sculp.

Caltha palustris L.

RANUNCULACEAE

This very elegant plant has rounded, fleshy leaves of fine, glossy green and flowers of a magnificent, bright orangey yellow. As its name suggests, the marsh marigold prefers to grow in very damp situations, even with its roots in the water. It is found both on marshland plains (though rarely in the Mediterranean region) and along the edges of mountain streams and lakes, coming into bloom shortly after the snows have melted. It is also cultivated as an ornamental in ponds and lakes.

The fact it often grows close to poplars (from the Latin *populus*), which also enjoy damp situations, led Tournefort (1700) to call the plant *populago*, a name constructed on the same principle as *tussilago* (coltsfoot), *medicago* (lucerne) and *plantago* (plantain).

The marsh marigold is also known in French as *souci d'eau* (water marigold). One of the ancient names for the marigold, *Caltha*, was simply an allusion to the colour of the flower, marigolds being well-known members of the composite family, opening with the sun, following its course and closing when the sun disappears from sight (from the Latin *solsequia*, 'that follows the sun').

The word ranunculus is also associated with damp environments. It means 'little frog' – from the Latin *rana* (frog), hence the name 'rainette frog', and from *culus* (small). In fact most ranunculuses grow in places where frogs also like to live. In France, some ranunculuses are even known familiarly as *grenouillettes* (little frogs).

The ranunculus familiar to everyone is the buttercup. Children used to hold a buttercup under their friend's chin to find out whether or not they liked butter. But behind this tradition lies a highly toxic plant with a formidable poison! The Romans would poison their prisoners with the fresh juice of a ranunculus – the *sardonia* – also known as the Sardinian herb. In death the dying person's mouth would contort into a fixed grimace or rictus that Ambroise Paré later termed *ris sardonic* or *ris sardinien* (sardonic or Sardinian smile). Influenced by the term 'sarcastic', this became the evil and sardonic laugh we associate, for example, with the terrifying witch in *Snow White*.

The harmful substances produced by this plant belong to the ranunculoside group, which split on drying and thus lose their toxicity. This is fortunate as animals often eat dried buttercups in their hay, whereas they avoid eating them fresh in their pastures.

In 1931, a doctor named Gessner drew attention to cases of poisoning 'due to the use of flower buds like capers, of leaves mixed into salads, and of flowers sometimes used to colour butter' (Fournier 1948). However, some recent cookbooks include recipes for marsh marigold buds preserved in vinegar for use as condiments!

Caltha palustris L. Engraving : Nicolas Robert
RMN, Louvre Chalcography Dept, Paris (modern print).

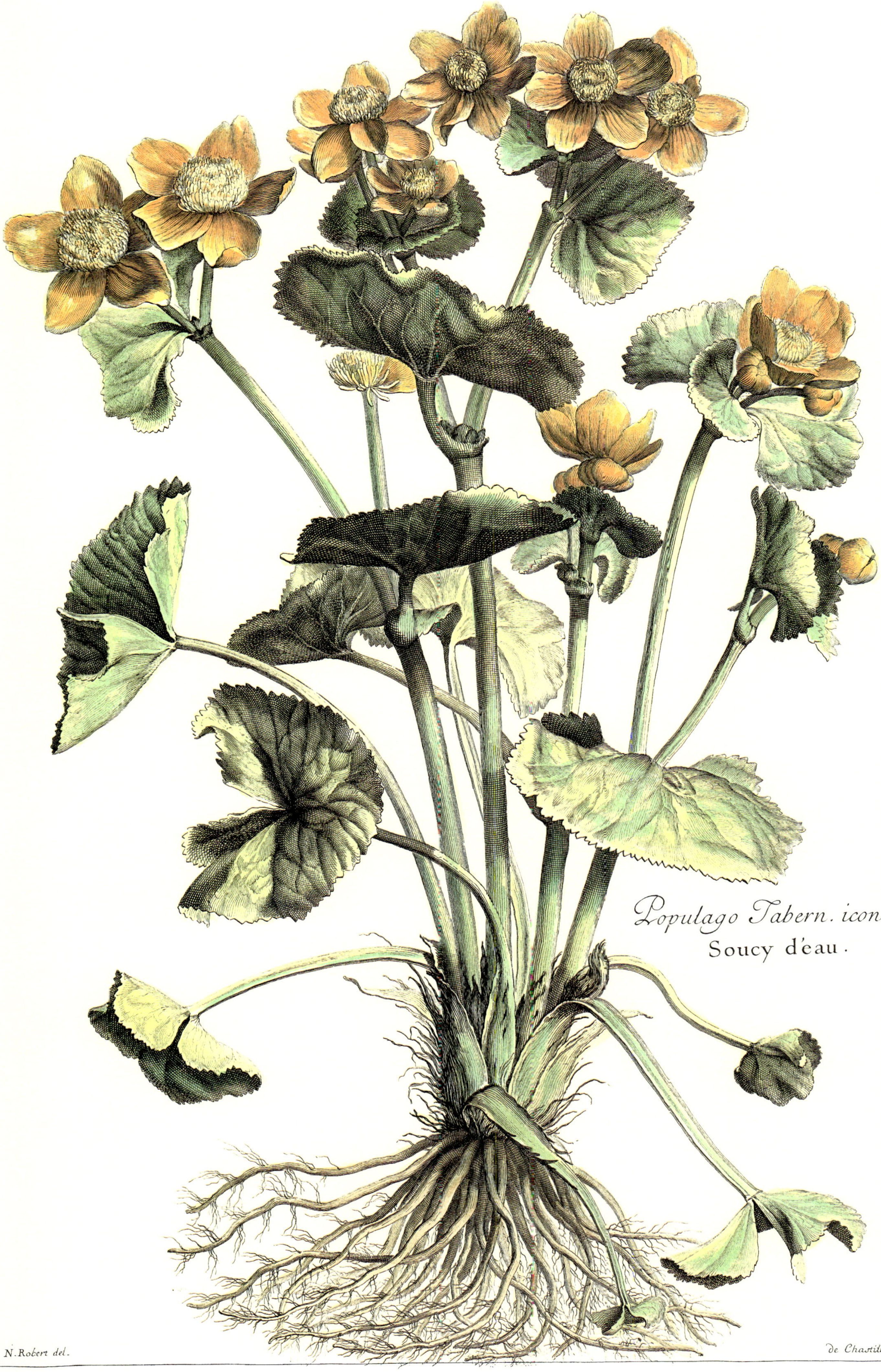

Populago Tabern. icon. 750.

Soucy d'eau.

N. Robert del. — de Chastillon sculp.

Eranthis hyemalis (L.) Salisb.

RANUNCULACEAE

Before becoming known as *Aconitum hyemale*, this plant was called *Aconit tue-loup, jaune, petit* (small, yellow wolfsbane) by Rembert Dodoens (1517–85), known as Dodon. Dodon was a Dutch botanist and physician to the emperors Maximilian II and Rudolph II. He is known in particular for his *History of Plants*.

Here is how Dodon describes the plant in a text quoted by Daléchamps (1615): 'It is a small herb that has scarcely any stem, apart from the tails of its flowers, which are short and barely a palm span in height, and though they are spindly, on each of them there is an enclosing collar-like single leaf: however this is jagged, like that of the *Napellus*, with large ray-like serrations, in the centre of which is a small, yellow flower, similar to that of ranunculuses, after which comes not a bud but three or four small, straight pods, like those of the *Lycoctonon*. Its root is fat and rounded ... and somewhat knotted, with a strong, acrid taste, and from it emerge the tails of the leaves. It flourishes in the gardens of Flanders, quite often in the month of January when the winter is mild, and because of this, people in that area have begun to call it *Aconitum hyemale* or *hybernum*. For it appears from the form of its leaves and of its pods, and from the pernicious property of this plant that it is a species of aconite. In fact Matthiol has it as a second species of *Aconitum lycoctonon*, or wolfsbane.'

Another botanist describes it as *Helleborus pumilus* ('dwarf' in Latin). Linnaeus (1707–78) confirms this as the genus but retains the species name *hyemale* ('winter' in Latin). The British botanist Richard Anthony Salisbury (1761–1829), author of several works including *Genera of Plants* (published after his death, in 1866), gave it its definitive name of *Eranthis hyemalis*, the binomial name by which it is still known.

The *Eranthis* is believed to have originated in the mountains of southern Europe, though where exactly is unknown. In France it is found mainly in the east, the centre, the Alps and the north-west, growing in bright forest clearings, under hedges, in thickets and on roadsides.

Like salsify, chicory and certain other plants, winter aconite flowers open in the morning when the weather is sunny, but last for one week only, becoming larger during their flowering period. It is a poisonous plant, like the hellebore, which has a disputed etymology. According to the botanist Gaston Bonnier (1934), the name comes from the Greek words *hélein* (to destroy) and *bora* (pasture). Fournier (1948) associates it with the Semitic word *helibar* or *helebar* (madness) – the eastern hellebore having enjoyed a reputation since ancient times as a cure for madness. The term has also been related to *Aliboron*, meaning universal remedy.

Eranthis hyemalis (L.) Salisb. Engraving : Nicolas Robert
National Museum of Natural History, Paris.

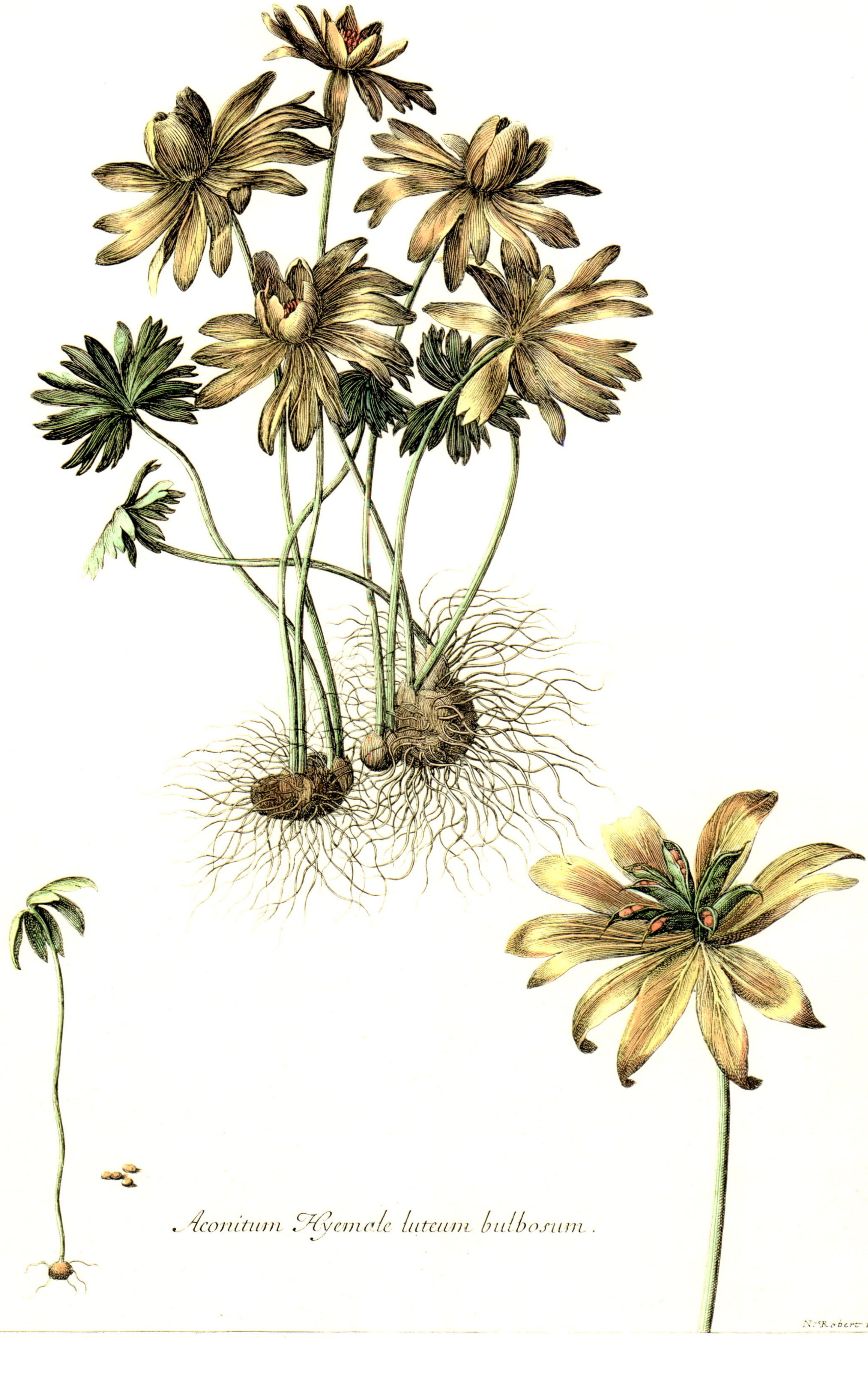

Aconitum Hyemale luteum bulbosum.

N. Robert in. f.

Anemone nemorosa L.

RANUNCULACEAE

The difference between the ranunculus and the anemone lies mainly in its flower. A ranunculus flower has five very distinct sepals and petals, while that of the anemone has no petals. What we find so striking are in fact the coloured, petaloid sepals. The number varies from five to nine, or even more. The wood anemone exists in a wide palette of colours, ranging from white that is pinkish on the outside, to pink, purple, violet and blue, as seen in the engraving. The wood anemone (known in French as *anémone Sylvie* or *grenouillette*) is a very common plant that colonises forest floors and damp meadows. When it blooms in spring it transforms the woods into magical places.

Daléchamps (1615) describes it thus: 'According to Fuchse, the white *grenouillette* which is smaller, also has two forms: one with white flowers which is the one also described by Dioscorides, and another with yellow flowers. The first produces three leaves, of a similar jagged form to other *grenouillettes*, and between these appear pale or white flowers: its root is spindly, twisted, with a great deal of hairy root below, which stretches deep into the ground. The second is different only in that its flowers are yellow. Dodon calls both *Ranunculus parvus* or *nemorensis*. They grow in woods, ditches and damp, shady places.'

Though reputed to be very acrid, vesicant, corrosive and toxic, this plant has nevertheless been used as a popular remedy for paralysis, scabies, alopecia, toothache and other ailments.

The name 'anemone' comes from the Greek *anemos* (wind) which has a Sanskrit root, *âni-ti* (to blow), as the plant moves at the slightest breath of wind. It is also related to the Latin words *animus*, *anima* (soul). The French name *Sylvie* comes from the Latin *silva* or *sylva* (forest) and *sylvaticus*, which evolved into *salvaticus* to produce the French word *sauvage* (wild).

The anemone is associated with one of the great plant myths – that of Adonis. According to this legend, which is Syrian in origin, Theias, king of Syria, had a daughter named Myrrha (or Smyrna) whom Aphrodite in her wrath compelled to commit incest with her father. With the aid of her nurse Hippolyta, she managed to deceive Theias and slept with him for twelve nights. But on the twelfth night, the king discovered his daughter's deceit. She fled, pursued by her father who was armed with a knife to kill her. Myrrha implored the gods to help her and she was transformed into a myrrh tree. Ten months later, the bark of the tree peeled back and opened, and an infant emerged who was given the name Adonis. Aphrodite, moved by the child's beauty, took him and gave him to Persephone, the goddess of the underworld, to raise. But as the child grew up, Persephone fell in love with him and did not want to return him. This dispute was settled by Zeus, who decided that Adonis should live for one third of the year with each of the two goddesses and the final third wherever he wished. He chose to spend it with Aphrodite, in a symbolic reflection of the life of plants, which spend two thirds of the year above ground and one third below. Following some dark episodes of love, jealousy and vengeance, of the kind in which Greek mythology abounds, the goddess and huntress Artemis sent a boar to attack Adonis and he was mortally wounded. According to the myth, from each drop of his blood an anemone grew.

Several other legends surround this myth: Myrrha's tears are said to be the origin of myrrh; red roses are said to have come from white roses – Aphrodite, running after her wounded love, pierced her foot with a rose thorn and her blood coloured the blooms. It is also said that red roses, the symbol of true love, are dedicated to her.

Anemone nemorosa L. Engraving : Louis de Chastillon
RMN, Louvre Chalcography Dept, Paris (modern print).

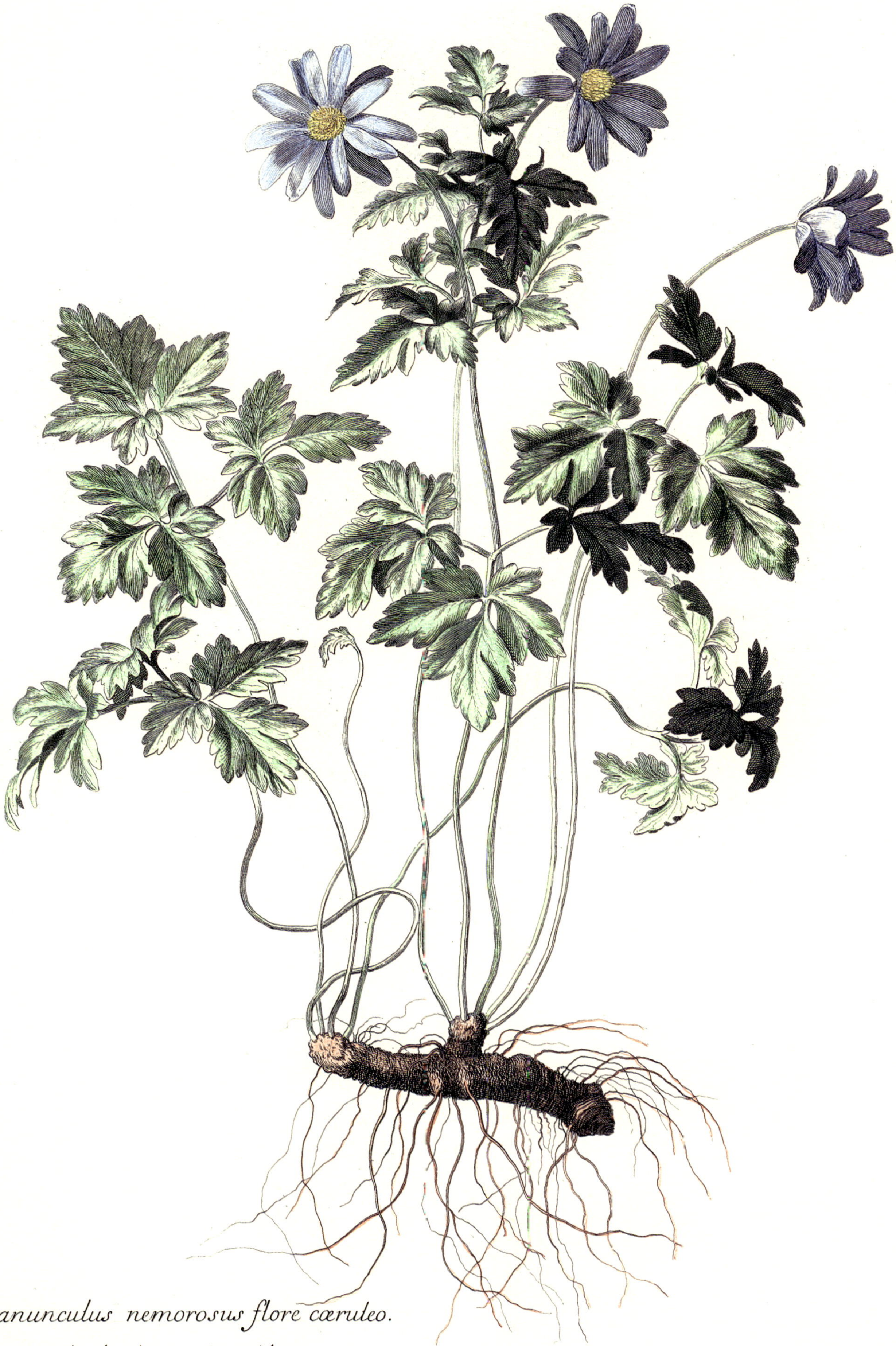

Ranunculus nemorosus flore cœruleo.

Ranuncule des bois a fleur bleuë.

de Chastillon sculp.

Trollius europaeus L.

RANUNCULACEAE

Despite its name, this plant has nothing whatever to do with trolls – neither those mischievous little figures of Scandinavian mythology nor the evil beings from the world of Tolkien. The name derives from the form of the flower which, if it ever opens, resembles a basin or bowl, and comes from the Latin *trulleus* (basin). It was given this name by Ruppius in 1718.

Heinrich Bernhard Rupp (1688–1719), also known as Ruppius, was a German botanist. He began studying medicine in 1704, continuing his studies in Jena and Leiden. Ruppius produced two volumes of flora, one devoted to the area around Jena, the other to Thuringia in central Germany. The genus *Ruppia* of the family Ruppiaceae was dedicated to Ruppius by the botanist Linnaeus (1707–78).

In the sixteenth century, this plant was known in France as the *grenouillette à fleurs rondes* (round-flowered ranunculus). This is how Daléchamps describes it in 1615: 'Its flower is also the colour of gold [...] but always remains like a bud and smells good. It grows on the mountains in Switzerland, especially on one known as the broken mountain. Gesnerius says that it is commonly known as *Trollius florem*. Lobel calls it *Ranunculus montanus alpinus glomeratus*, and says that it likes to grow in the mountains of the valleys of Aosta, Mende, the Auvergne, Germany and Switzerland'.

Known in English as the globeflower, this species is typically found in damp meadowlands and open mountain woodland. Though it may be abundant locally in certain areas, it is considered rare in France as a whole and is indeed a protected plant in several Swiss cantons. It can be found up to an altitude of 2,500 metres (8,200 ft) or occasionally even higher. A meadow filled with globeflowers, with snow-covered mountains in the background, is a stirring sight. However, alpine farmers are far from happy to see their pastures invaded by this poisonous plant shunned by animals. As with many ranunculuses, the protoanemonin it contains is dangerous when fresh but is transformed into non-toxic anemonin on drying. This means that it is safe for animals to eat hay containing it.

The globeflower bloom looks like a very large buttercup that has scarcely opened, the petaloid sepals remaining curved in on themselves. In this illustration the artist probably opened the flower to make it appear mature! Bees and other large insects are unable to enter it to drink nectar and ensure pollination takes place. In fact, it is a small fly – an anthomyiidae of the genus *Chiastocheta* – which, in a mutually beneficial plant-insect relationship, takes care of this. The insect enters the flower, pollinates it and lays its eggs. By thus providing its larvae something to feed on, and therefore grow, this fly has found that special little niche essential to its survival.

Trollius europaeus L. Engraving : Abraham Bosse
National Museum of Natural History, Paris.

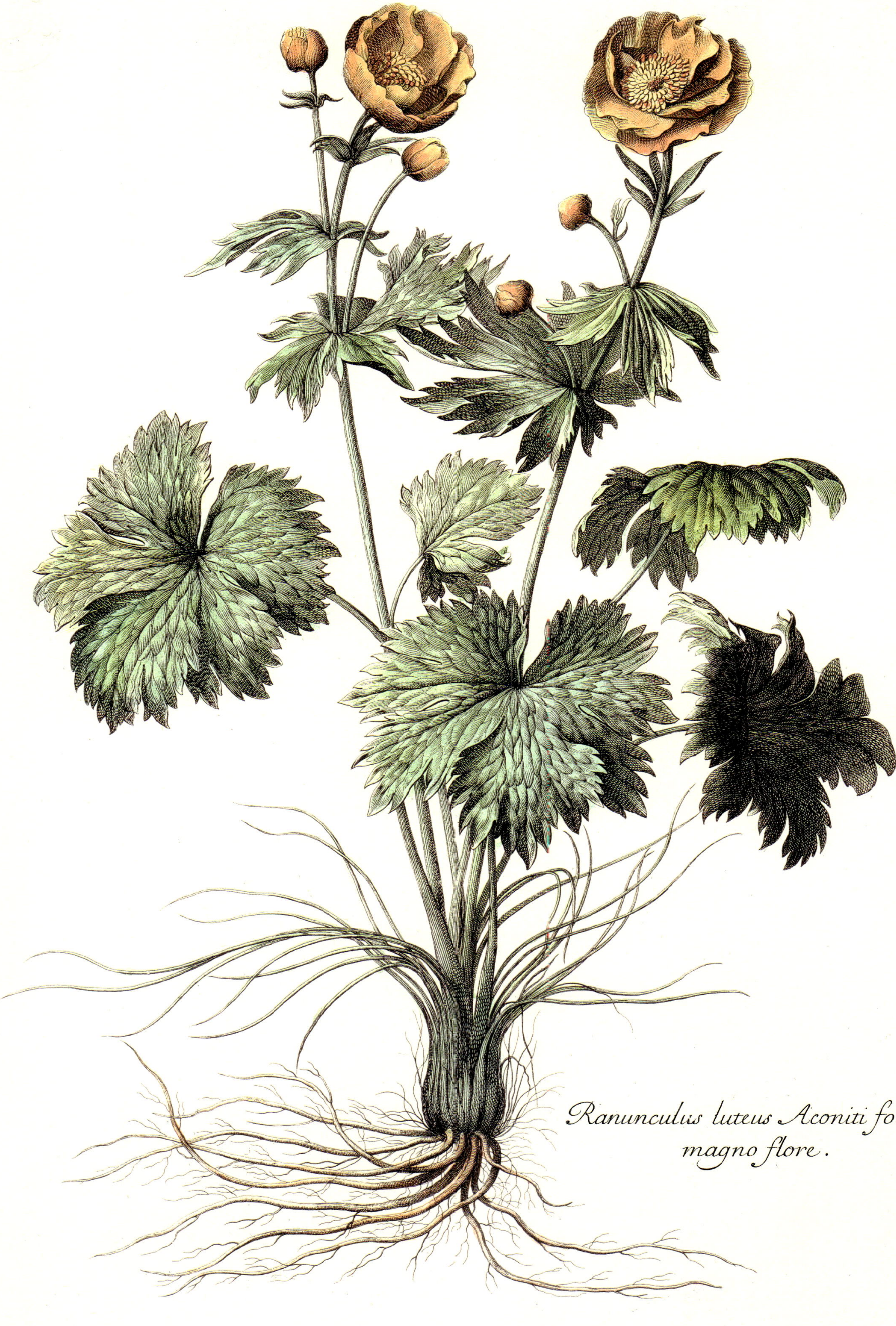

Ranunculus luteus Aconiti folio magno flore.

Bosse sculp.

Glaucium flavum Crantz

PAPAVERACEAE

Shown here under the name *Papaver corniculatum flore luteo*, this plant presents all the characteristics of *Glaucium flavum* (yellow horned poppy) in its foliage and arching, glabrous seed capsule, but the flower, which is normally yellow, is here shown as red. A red-flowered *Glaucium corniculatum* (L.) J.H. exists, but its foliage is far more serrated, in fact deeply pinnatisect, and its capsule is straight with short, stiff bristles. The illustrator – who has made an excellent drawing of the plant – has, however, painted the flowers red, although the word *luteo* means yellow!

The foliage of both *Glaucium* are glaucous, hence their name. These two species grow in the southern regions of France, but *G. flavum* extends over a wider area, from the west and the centre of the country as far north as the Pas-de-Calais. Both are found where the soil has been disturbed: *G. flavum* more in areas of rocks, sand and rubble and *G. corniculatum* in fields and on hillsides.

Both these *Glaucium* were known to the Ancients. Daléchamps (1615) devotes two pages to them: 'Matthiol says that they are found in force on the shores of Siena ... and also along the Adriatic Sea. Clusius recognises this poppy as the horned poppy, as does Dodon who has it as the great horned poppy ... Its flowers are an ochre yellow, after which instead of a testa comes a long, small, slender, curved horn, or pod, containing black seed ... In addition to this horned poppy, which is common, Clusius and Dodon observed two other species, the first of which has jagged leaves like rocket, with deep serrations ... Clusius describes it thus: and also this species of poppy has been picked in Languedoc among the oats, close to the village of Latez, along the river Lez ... Furthermore, the root of the horned poppy, as Dioscorides tells us, cooked in water until it has reduced by half and taken as a beverage will help to cure sciatica and liver disorders. It is beneficial for those who have dirty, cloudy urine. An acetabulum measure of the seed, taken with honeyed water, will act as a light purge for the stomach. Its flowers and leaves, applied with oil, will remove the scabs on ulcers. [...] but they should not be applied after the ulcers have been cleansed, because they are so detergent that they consume living flesh ... They will remove specks and dirt from the eyes of small livestock, if rubbed on them ... The horned poppy, when eaten or taken as a beverage, produces the same effects as opium ...'.

Many members of the poppy family are used for their medicinal properties. In addition to the well-known *papaver somniferum* (opium poppy), there is the field poppy whose petals have cough suppressant, antispasmodic and neurosedative qualities; the celandine, known for its orangey-yellow latex that is applied to warts and for its antispasmodic action in biliary tract disorders; and *Eschscholzia* (the California poppy) which is a sedative and hypnotic. In fact, the poppy family as a whole provides many treatments for stress and nervous disorders.

Glaucium flavum Crantz. Engraving: Nicolas Robert
RMN, Louvre Chalcography Dept, Paris (modern print).

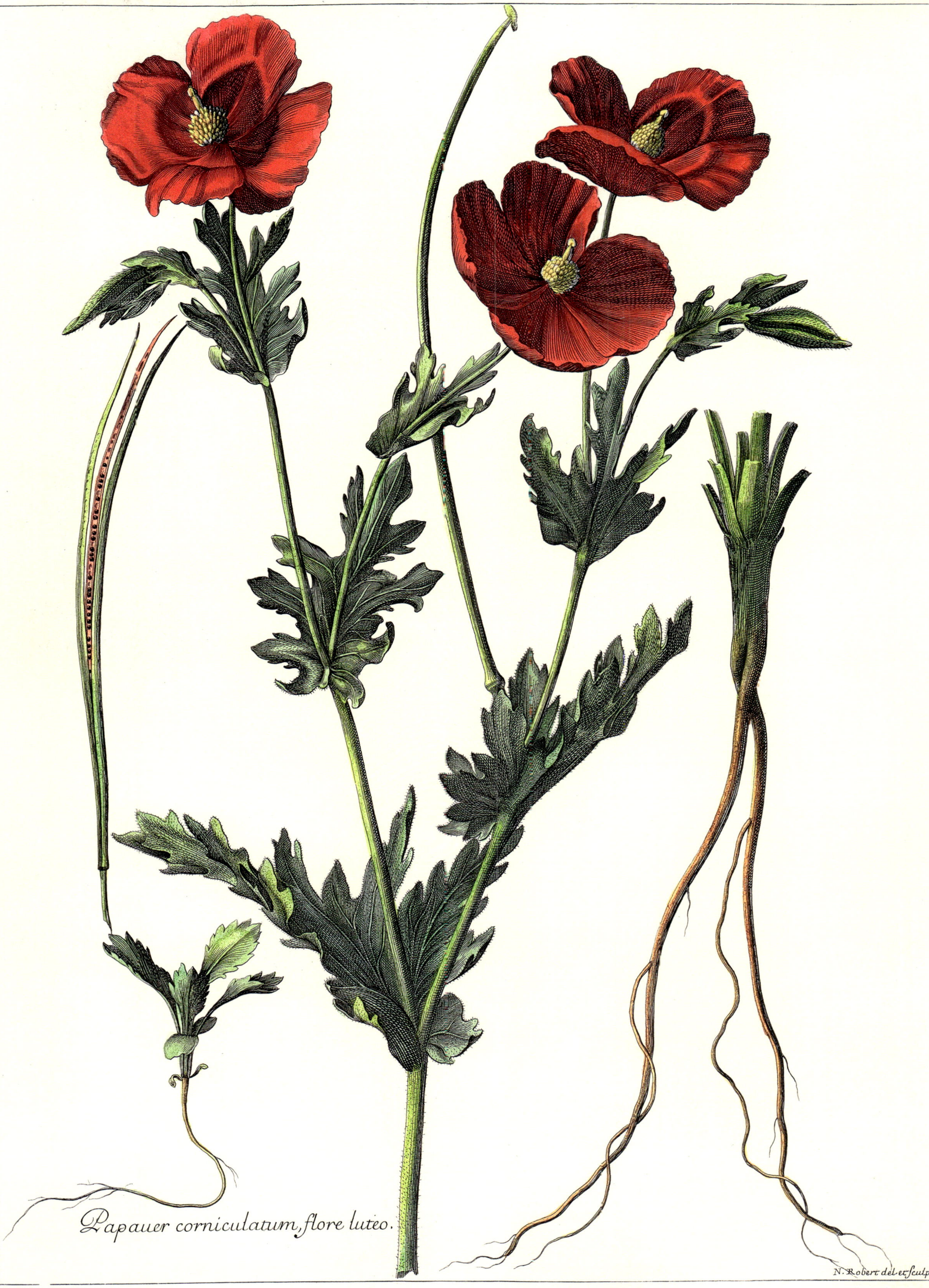

Papauer corniculatum, flore luteo.

N. Robert del. et sculp.

Melocactus sp.

CACTACEAE

This was probably one of the very first cactuses seen by the discoverers of the New World and is a highly original plant. The oldest representation of this species appears in a work by Mathias Lobel, known as Lobelius, co-written with Pierre Pena and entitled *Stirpium adversaria nova* (1570).

The overall appearance of the *Melocactus* is remarkable. The part where photosynthesis takes place, known as the 'body', is very much what we imagine a cactus to be, that is to say a ribbed succulent covered with spines. What is different about the *Melocactus* is that it has two distinct parts: the vegetative part, the 'cactus' which seems to come to an abrupt end, and the reproductive part, the cephalium, which stands erect and can be cylindrical or globular, depending on the species. This part is formed of small tubercles covered in hair like wool which are small, embedded flowers, only the corolla of which is visible. Depending on the age of the plant and the species, this can be as much as a metre (3 ft) tall and many people see it as phallic.

Leaving aside the originality of its flowers, this cactus is, like other members of the family cactaceae, very well adapted to the xeric conditions of its environment: its long roots enable it to search for water at a distance and give it good anchorage in the soil, which is essential when the plant is topped by such an imposing floral appendage.

According to Gérard Delanoy, a *Melocactus* expert, certain species are self-fertile, while others require hummingbirds, insects and even lizards for their pollination. The seeds of the species *M. citispinus* and *M. macracanthos* are dispersed by various insects, birds and lizards, including *Chemidophorus murinus ruthveni*, while iguanids perform this function for the Brazilian *Melocactus*.

Amazingly, this cactus is mentioned as early as 1615, by Daléchamps: 'We must not forget to include here a kind of foreign thistle, both for its rarity and for its beauty: we have included here the portrait and description made by Pena. This thistle, he says, grows naturally in the Western Islands of the New World, particularly on one known as Santa Margarita, from whence English mariners brought it to Morgan the Apothecary in London. It appears that this is one of nature's combination plant, composed of a melon and a thistle, and for this reason it is called *Echinomelocactos* or *Melocarduus Echinatus*. It is round and made in the form of a spinning top and has fourteen grooves and thick ribs ... and forms a kind of flattened plate from which emerges, as with the Artichoke, a silky wad of hair ... From these grooves, emerge sharp thorns, like those of a porcupine or a hedgehog, and as hard as horn ... But the pulp within is white, fatty, flaccid and soft, something like the flesh of a melon or animal fat, with an insipid taste and a little sourish but not unpleasant ... Those who have seen it in situ, such as sailors and others say that this thistle grows in sandy places near the sea. When it is ripe it is the size of a medium or large melon. It grows flat against the ground, with no leaves visible ...'

More than 2,000 species of *Melocactus* grow along the coasts of Mexico and Central America, in the West Indies, Ecuador, Peru, Venezuela and Brazil.

Melocactus sp. Engraving : Louis de Chastillon
RMN, Louvre Chalcography Dept, Paris (modern print).

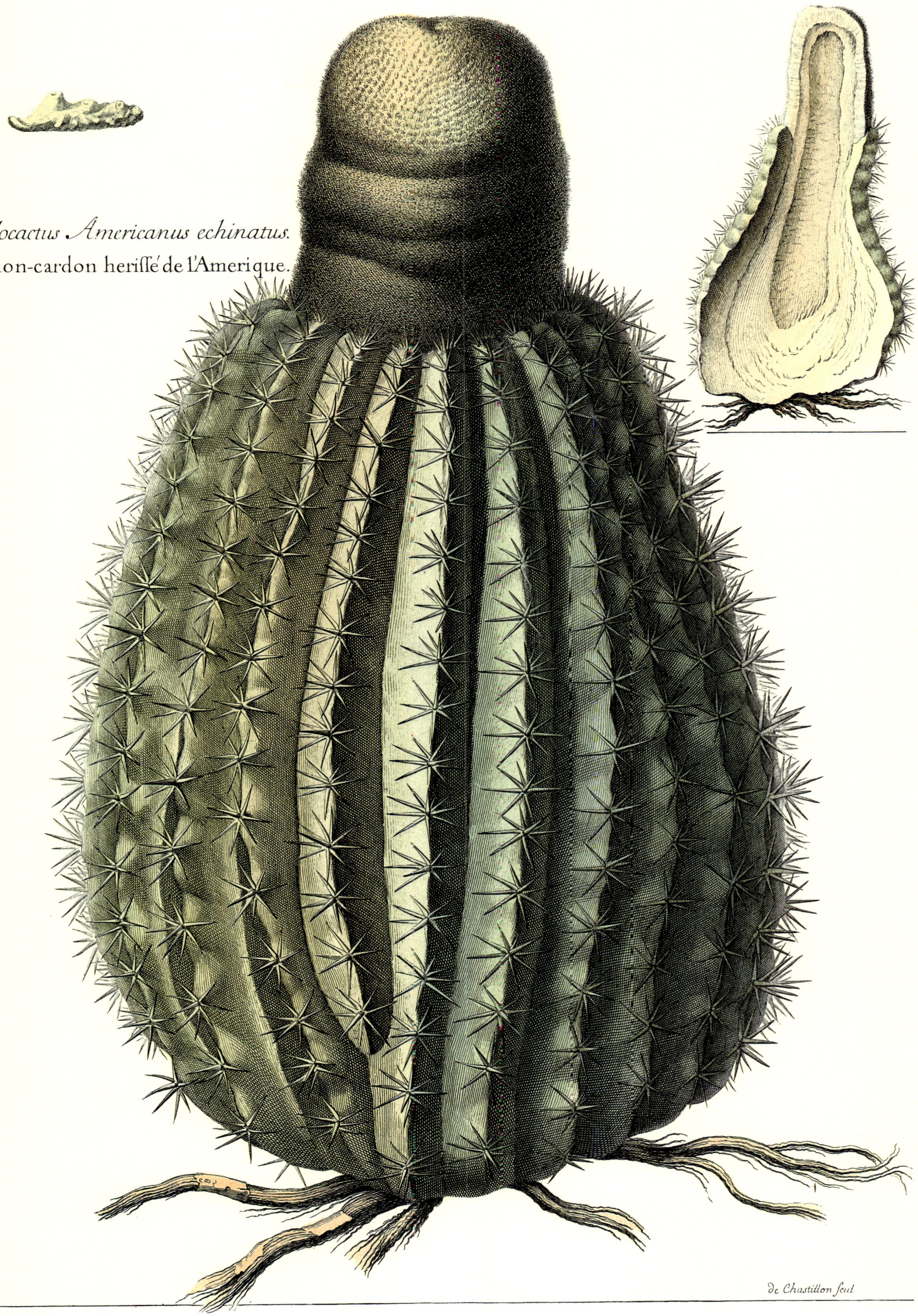

Melocactus Americanus echinatus.
Melon-cardon herissé de l'Amerique.

de Chastillon sculp.

Lychnis sp.

CARYOPHYLLACEAE

We have very little information about this *Lychnis* and there is no trace of it in either Caspar Bauhin's *Theatri Botanici* (1658) or Tournefort's *Institutiones rei herbariae* (1700), nor is it found in Linnaeus's *Species plantarum* (1753). So it comes to us under the splendid name of *Lychnis sicula glabra pseudomelanthii facies*, the yellow *lychnis* of Sicily. Colour apart, it could be mistaken for *Silene coeli-rosa* (L.) Godron, an adventitious species found in Lebanon, Libya and Syria, with a characteristic calyx tapering at the base. On French territory it is localised only in Corsica and appears on the list of protected plants in France.

Lychnis comes from the Greek *lychnos*, meaning lamp, the wicks of which were once made from the soft stems of the species *coronaria*, the rose campion. This species, which appears to be native across an area extending from southern Europe to western Asia is highly ornamental with its silvery, downy leaves and bright purple flowers. In France it is part of the spontaneous flora but is also cultivated in gardens and has been used to produce some very attractive varieties. It is known by various names in France including *coquelourde des jardins*, *coucourde*, *passerose*, *passefleur* and *œillet de Dieu*, but in English is simply the rose campion.

The name *coquelourde* is also used to refer to the *Anemone pulsatilla*, a poisonous plant much used in allopathic and homeopathic medicine and known by a number of different French common names including *coqueret*, *coquerelle*, *coquerette*, and in English by the names pasque flower, windflower and meadow anemone.

The term *coquelourde* is associated with a person who is vain and foolish. This may be due to a certain similarity between the feathery plumes on the fruit of the pulsatilla and a cockerel's comb – the intelligence and behaviour of cockerels leaving something to be desired in the popular imagination. Or perhaps because the anemone, whose name comes from the Greek word *anemos* (wind) moves easily in the lightest breeze. Or maybe because the fruit of the *Silene* are hollow and swollen like the sides of an empty wineskin – a reference to the enormous belly of Silenus, Bacchus's drinking companion.

The longstanding dispute over *Lychnis* and *Silene* was brought to an abrupt end by Kerguélen. In his *Index synonymique de la flore de France* (1993), all *Lychnis* are called *Silene*. Thus, *Lychnis coronaria* (L.) Desr. has become *Silene coronaria* (L.) Clairv. The same may also be true of the yellow lychnis shown here.

Lychnis sp. Engraving : Nicolas Robert
RMN, Louvre Chalcography Dept, Paris (modern print).

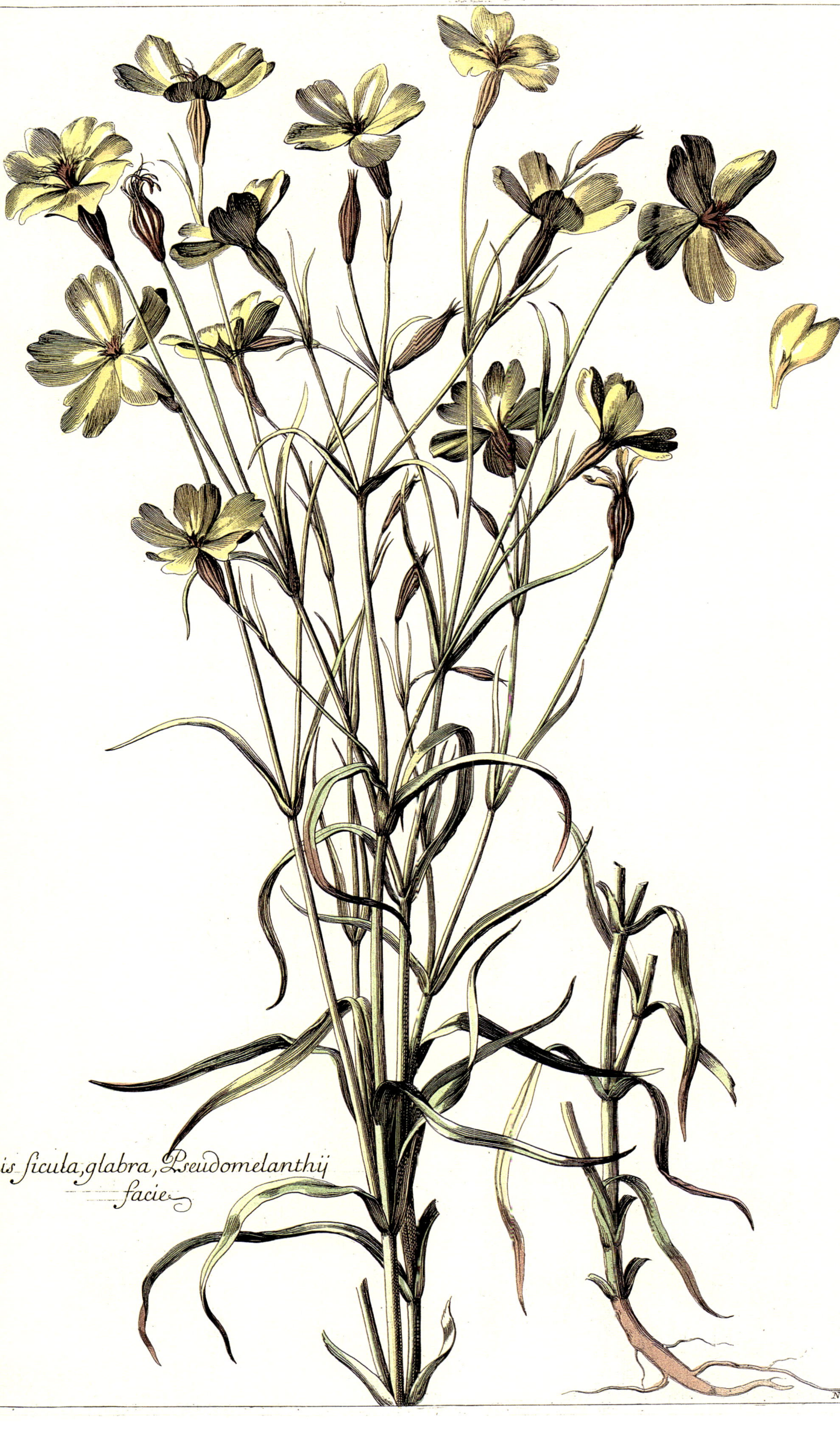

Lychnis sicula, glabra, Pseudomelanthij facie

N. Robert del. et Sculp.

Emex spinosa (L.) Campdera

POLYGONACEAE

In 1620, Caspar Bauhin named this plant, with its beetroot-like leaves, *Beta cretica semine aculeato* and placed it in the *Beta minor* group.

Emex spinosa is almost non-existent in France and it is therefore very surprising that it appears in Fournier (1946) and Kerguélen (1993), under the name *Vibo spinosa* (L.) Moench. Known for a time as *Rumex spinosus* L., it was classified as a sorrel. Fournier believed it to be very rare and refers to it only as an adventitious species in Corsica, in the area around Bastia.

Flora Europea (1976) mentions it in the Mediterranean countries of Spain, Portugal, Italy, Sicily, Greece, Tunisia, Algeria, and Morocco, where it grows happily in disturbed environments, fields, sandy spots and roadsides. (Another species, *Emex australis* Steinheil, is found in similar environments in warmer regions.) In some areas it is considered to be a field weed and is beneficial only to certain birds, such as the spotted sandgrouse. In Iraq, this granivorous bird, resembling a partridge and living in desert environments, consumes large quantities of its seed. Some countries are keeping a close watch on the spread of *Emex*, Australia in particular which has experienced serious plant invasions with *Opuntias*, rush skeletonweed, St John's wort, purple viper's bugloss and others.

Emex – known in Arabic as *bezzoulet naja* in Tunisia, as *el hanzab* in Libya, as *hummayda* and *silq* in Morocco (names which include the sorrels) – has the appetising appearance of a small, wild beetroot. In Arab countries the root is eaten raw and the young, tender leaves are used to prepare *idam*, a kind of rich sauce. The roots and leaves are used medicinally: transformed into a broth they are traditionally used to treat gastritis and gastric ulcers.

In actual fact, the only spiny feature of the plant are the sepals, which become hardened after flowering. They form a tough tube around the fruit with their tips curved into hook-like spines that facilitate dispersal by sticking to the fur of animals (epizoochoria) – a system also used by other plants such as agrimony, wood avens and burdock.

Burdock made Georges de Mestral a rich man. One autumn day when he was out walking, his dog disappeared into an area infested with burdock. When the animal returned it looked more like some sort of plant monster than a dog and Mestral spent a long time removing a vast number of spiky, hooked burrs. While doing this he was struck by the extraordinary gripping power of the plant's seed heads. The Swiss engineer decided to reproduce the technique using two types of synthetic fabric, one made of fine loops but smooth like velvet and the other made of flexible hooks. When brought together the two fabrics 'gripped' each other brilliantly and in 1951 he filed a patent under the name Velcro (Vel from *velours* – velvet – and Cro from *crochet* – the French word for hook).

Emex spinosa (L.) Campdera. Engraving: Nicolas Robert
National Museum of Natural History, Paris.

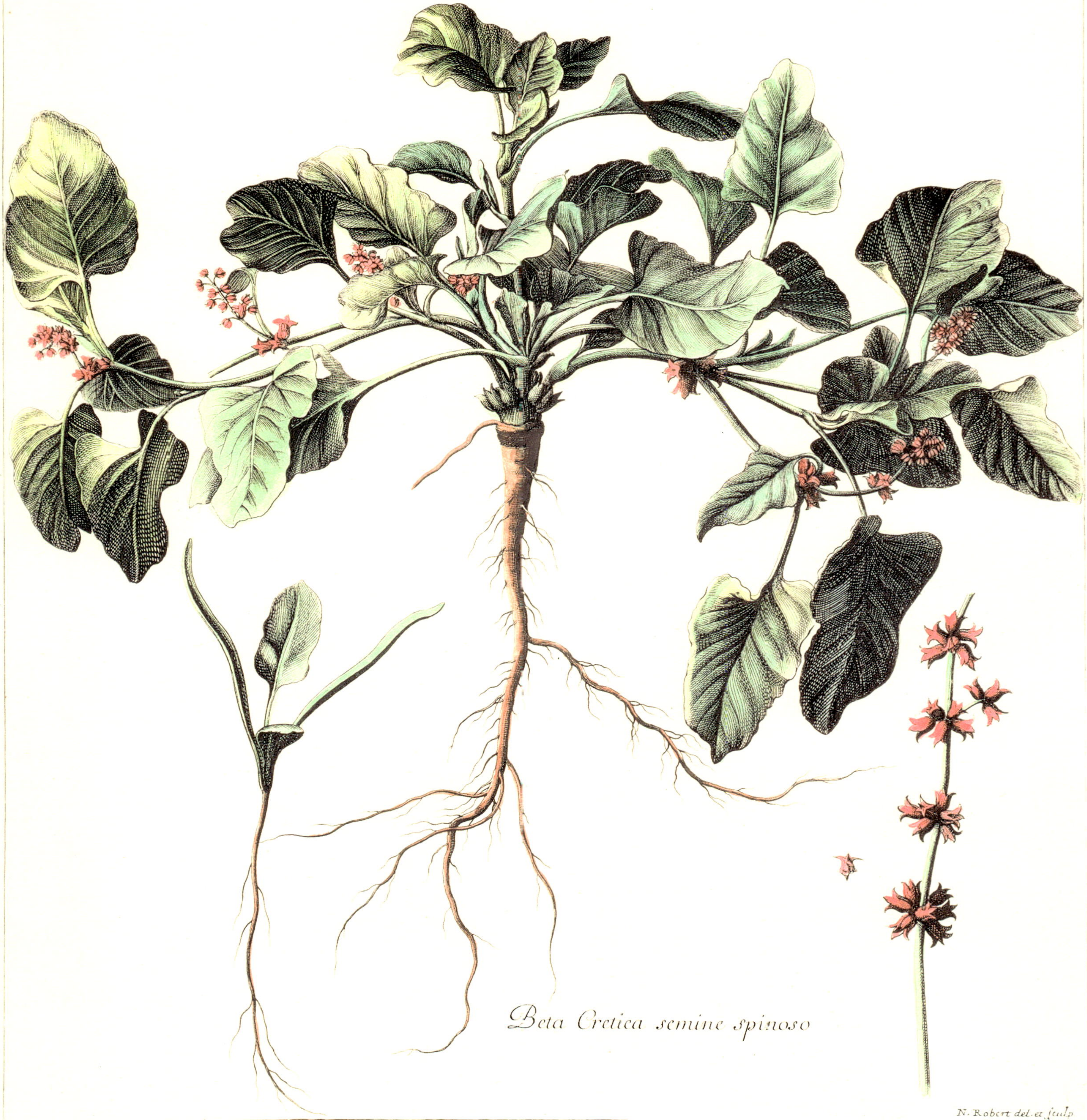

Beta Cretica semine spinoso

N. Robert del. et sculp

Gossypium sp.

MALVACEAE

Cotton is very well known for the thread and fabric it produces, and for its many clothing and household uses. Yet most Europeans are unfamiliar with the plant from which the fibre is harvested, the exception being those living in the more southerly areas of Europe – in Crete, Greece, Sicily, the south of Spain and Italy, Romania and the former Yugoslavia. In these countries, the cotton plant is still cultivated in some areas and has become naturalised in places. According to *Flora Europea* (1976), the four main species currently cultivated globally for cotton production can be found there: *Gossypium arboreum* L. and *G. herbaceum* L., both native to Asia; *G. barbadense* L., native to Peru; and *G. hirsutum* L., which is native to Central America.

Various parts of the plant are used: the fibre (the cotton), the linters (the soft cotton fibre that adheres to the seed after ginning) and the seed itself, which is rich in oil and protein. Once the toxic principle (gossypiol) has been removed from the cottonseed, cotton oil, the fifth most important food oil, is used, among other things, for frying and in the manufacture of cakes, margarine, soap and even certain lubricants. After the oil has been extracted, the remaining seedcake, which is rich in protein, is fed to cattle.

Cotton has been known throughout the Mediterranean basin since Antiquity. In the fourth century BC, Theophrastus refers to 'trees on the island of Tyle which bear wool and which have leaves like vines. The wool is enclosed in a fruit the size of an apple and when it is ripe the cotton is removed and fabric is made from it that can be either costly or inexpensive. This is also done in India and Arabia'.

Daléchamps (1615) quotes Pliny the Elder (23–79 AD): 'Pliny says that in upper Egypt on the border with Arabia, there grows a plant that some call *Gossipium* or *Xylon* and the fabric that is made from it *Xylina*. This plant is small and bears a fruit similar to the filbert,which has a cotton wad inside that is spun, and there is no wool in the world that is whiter or more delicate. The priests of Egypt have their robes made of it, to set them apart.' He adds, 'cotton does not only grow willingly in Egypt and India but nowadays it is also sown in Cyprus, Crete and Malta and in Sicily in great abundance and even in Calabria, in Puglia and other regions. The cotton gets hot and dries out and when burned has the marvellous virtue of stemming the flow of blood from wounds. Its seed, according to Serapion, is warm and moist. The seed pith is very good for coughs and for those who have breathing difficulties. It increases sperm production and consequently stimulates lovemaking. An oil is extracted from it that is very effective at removing face freckles and other bodily marks.'

Because of its economic and social importance, cotton has had a significant impact on society in terms of agriculture, trade, shipping and industry, but also in more reprehensible ways such as the slave trade, wars, revolts, ecological and humanitarian catastrophes and injustices of every kind. The official history of the slave trade begins on 8 January 1454, the date on which Pope Nicholas V gave the king of Portugal permission to practise this trade. In France the 'black code', an unbearable document to read, was promulgated by King Louis XIV in 1685, on the advice of Colbert; it authorises the slave trade in the colonies in order to further commercial domination by France and thus French global hegemony.

It is interesting to note that, strictly speaking, the term 'cotton lingerie' is incorrect. 'Lingerie' comes from the French word *linge* which originally only referred to linen goods and which was in fact made from flax.

Gossypium sp. Engraving : Abraham Bosse
National Museum of Natural History, Paris.

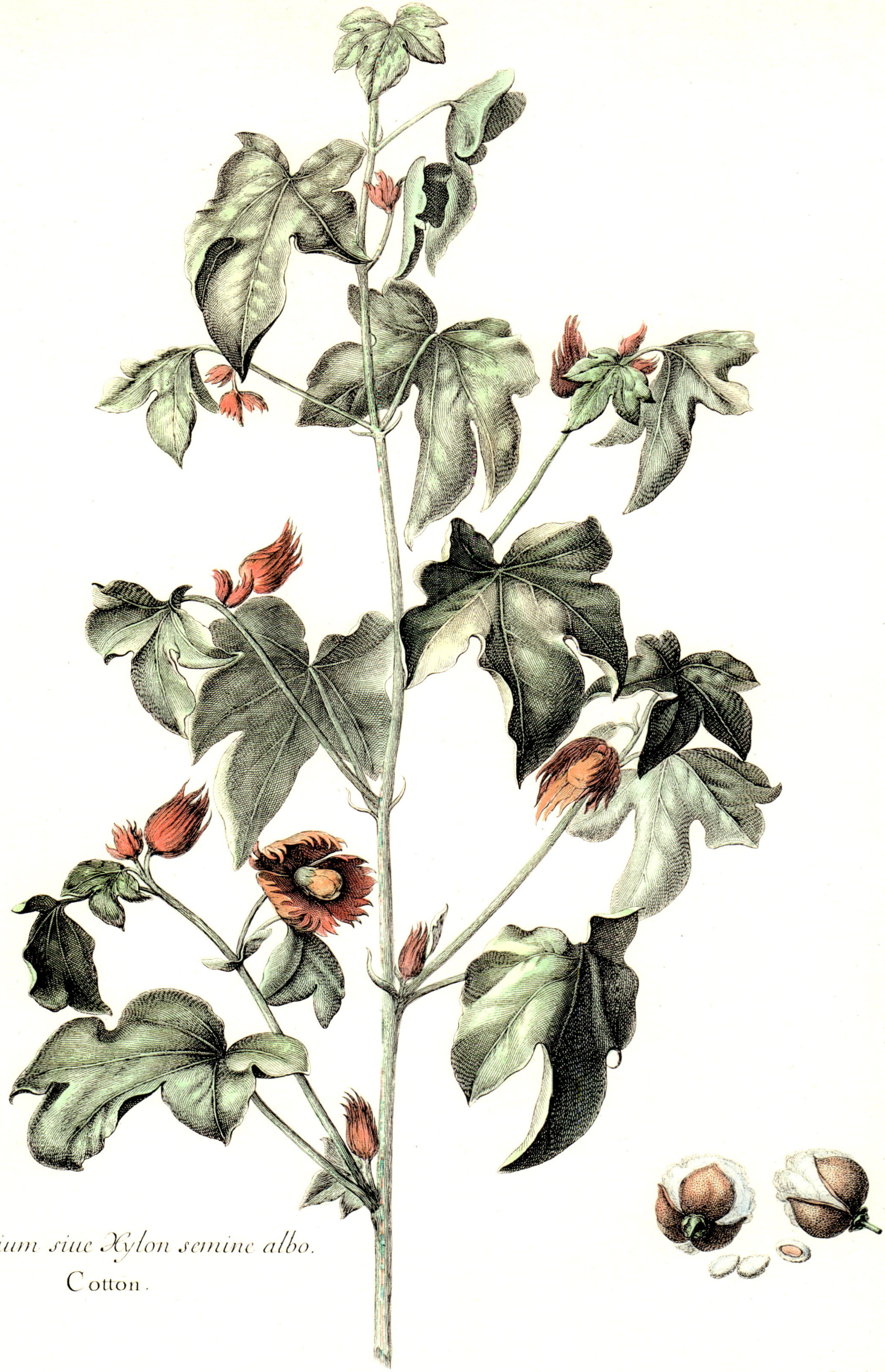

Gossipium siue Xylon semine albo.
Cotton.

A.Bosse sculp.

Urtica pilulifera L., Urtica sp.

URTICACEAE

The Roman nettle is similar in appearance to the common nettle (*Urtica dioica* L.) but is distinguished by bearing clusters of female flowers with rounded heads between ½ and 1 cm ($^{1}/_{5}$ to $^{2}/_{5}$ in) in diameter. It grows both in wasteland areas and in soils rich in nitrogen but in France is found only in the south and in Corsica.

It is less common than the small annual nettle (*Urtica urens* L.) and the common nettle with which we are all familiar, having stung ourselves at least once in our life: a nettle sting carries with it both highly aggressive formic acid and important chemical agents such as serotonin, histamine and acetylcholine. Our body reacts by quickly sending lymph to neutralise the aggression and heal the wound, creating a small blister.

The genus name is interesting in its own right and that of the species name, *di-oica*, means that the plant possesses two 'houses', a female and a male. There are in fact nettles that are only male and others that are only female. The term *oikos* is the origin of *oikos logos*: ecology – the science of the home extended to nature.

In the past, popular medicine advocated scourging rheumatic areas of the body with fresh nettle stems to 'bring blood' to the affected parts. This form of medication still has some brave adherents but others have opted for plant-based anti-inflammatories or more radical chemical treatments that are not without their own dangers.

Nettles have a number of medicinal properties: they contain a natural lactogenic hormone, secretin, which is why it is beneficial to breastfeeding mothers. In some Nordic countries nettles are grown extensively and used as dry fodder for dairy cows. Another substance, stigmastenone, found in the rhizomes of nettles, is recommended in cases of prostatic adenoma. A lectin, *Urtica Dioica Agglutinine* (UDA) is being studied for use in the treatment of AIDS – it strengthens the immune system and contains silica which helps remineralisation and is useful in fixing calcium.

The nettle was also used as a textile plant in many countries, such as Germany, for the manufacture of clothing and various fabrics.

Also, many gardeners leave nettles to soak in water to produce a form of liquid manure which is effective at warding off certain unwanted insects and also contains nitrogen, which stimulates plant growth. It is a pity that the nettle should so often be considered a weed when it has so many useful qualities.

Urtica pilulifera L., *Urtica* sp. Engraving : Nicolas Robert
National Museum of Natural History, Paris.

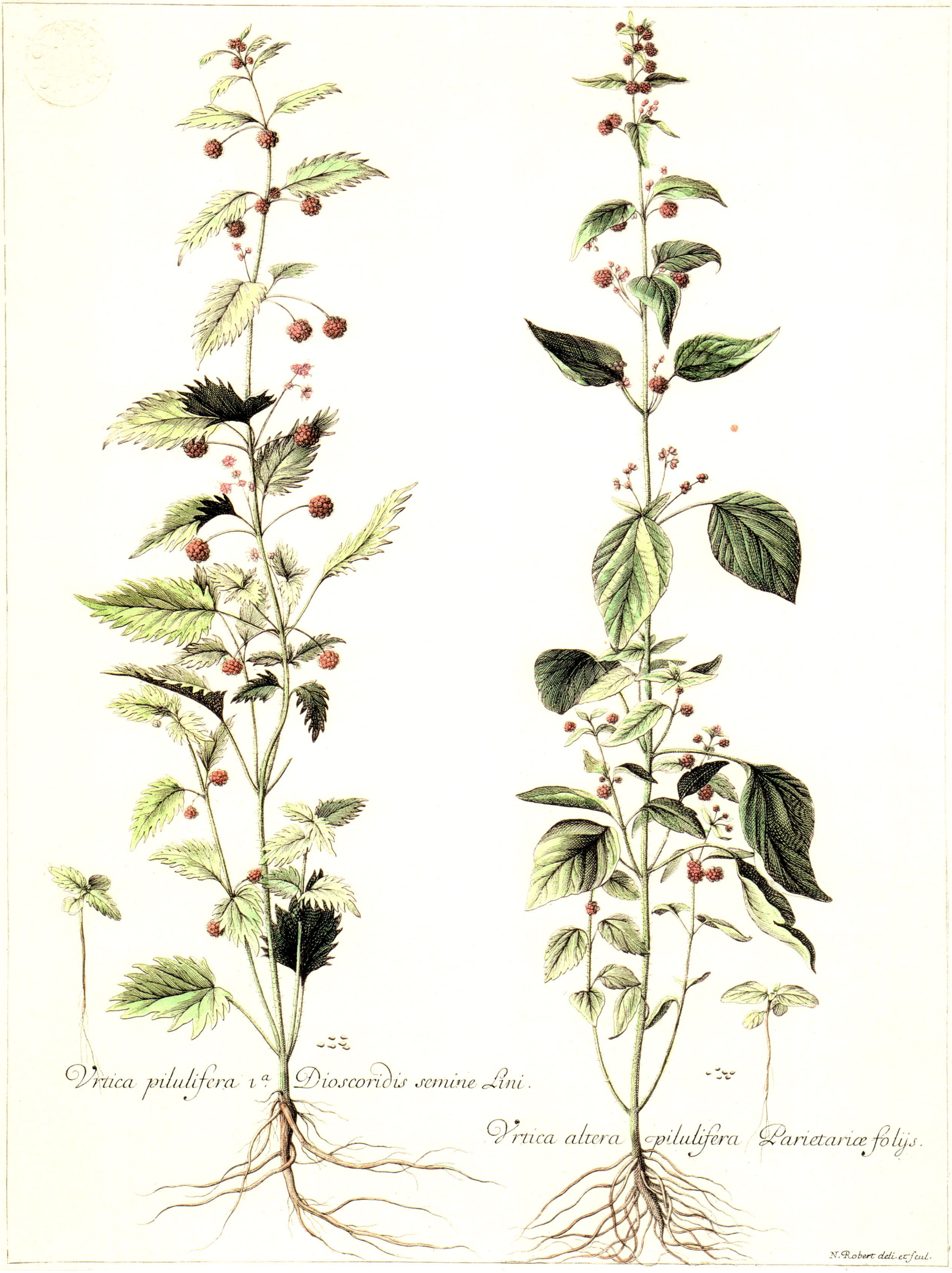
Vrtica pilulifera 1.a Dioscoridis semine Lini.
Vrtica altera pilulifera Parietariæ folijs.
N. Robert deli. et sculp.

Passiflora coerulea L.

PASSIFLORACEAE

It would have been difficult for Linnaeus back in 1735 to give any other name to this plant when, according to tradition, its extraordinary flowers were said to represent the instruments used in the crucifixion of Jesus: the crown of thorns is represented by the corona filaments; the nails of the cross by the stigma; and the wounds of Christ by the five stamens. Spanish monks also called it the 'flower of the five wounds'.

The Spanish explorer Pedro Cieza de León gave the first description of the plant and its flower in 1553, in Colombia. However, the name *Flor Passionis* is commonly attributed to a Jesuit monk, Jacomo Bosio, who met a Mexican monk called Emmanuel de Villegas who had in his possession drawings of the extraordinary flower.

Nicolas Monardes (1493–1588), a physician and botanist, describes it in his *Historia medicinal de las cosas que traen de nuestras Indias occidentales*, a work published in three parts (1565, 1571 and 1574) and translated into French by Charles de l'Écluse (1526–1609). Monardes, a businessman and friend of the viceroy of Mexico, played an active role in the introduction and dissemination of plants such as the pineapple, maize, coca leaves, sarsaparilla, jalap bark, peanuts and others, including guaiacum, which was used to treat syphilis, a great scourge at the time. He also produced the first illustration of the tobacco plant. Of the passionflower, Monardes wrote: 'The Aztecs used to call it *coanenepelli*, from *coatl* (serpent) and *nenepelli* (language), as they used it as an antidote to snake bites, perhaps because of the ... impressive form of its flower which resembles the open mouth of a snake.'

The Aztecs used several species of passionflower medicinally, for their calming properties. And the native indians already enjoyed the fruit and cultivated certain species such as *edulis* and *ligularis*. *P. coerulea*, the fruit of which is inedible, is often cultivated as an ornamental and trained over arbours, pergolas, or dead trees, but it can prove invasive.

In the indigenous culture, everyone was treated with plants. Hernán Cortés, who was healed of his wounds by the Aztecs, brought the most knowledgeable of them to Tlaltelolco to pass on their knowledge of medicinal plants. This led, in 1552, to the first handbook of herbal medicines, written and illustrated by Martinus de la Cruz but known as the 'Badianus manuscript' after Juannes Badianus who translated it into Latin.

Passiflora coerulea L. Engraving: Nicolas Robert
RMN, Louvre Chalcography Dept, Paris (modern print).

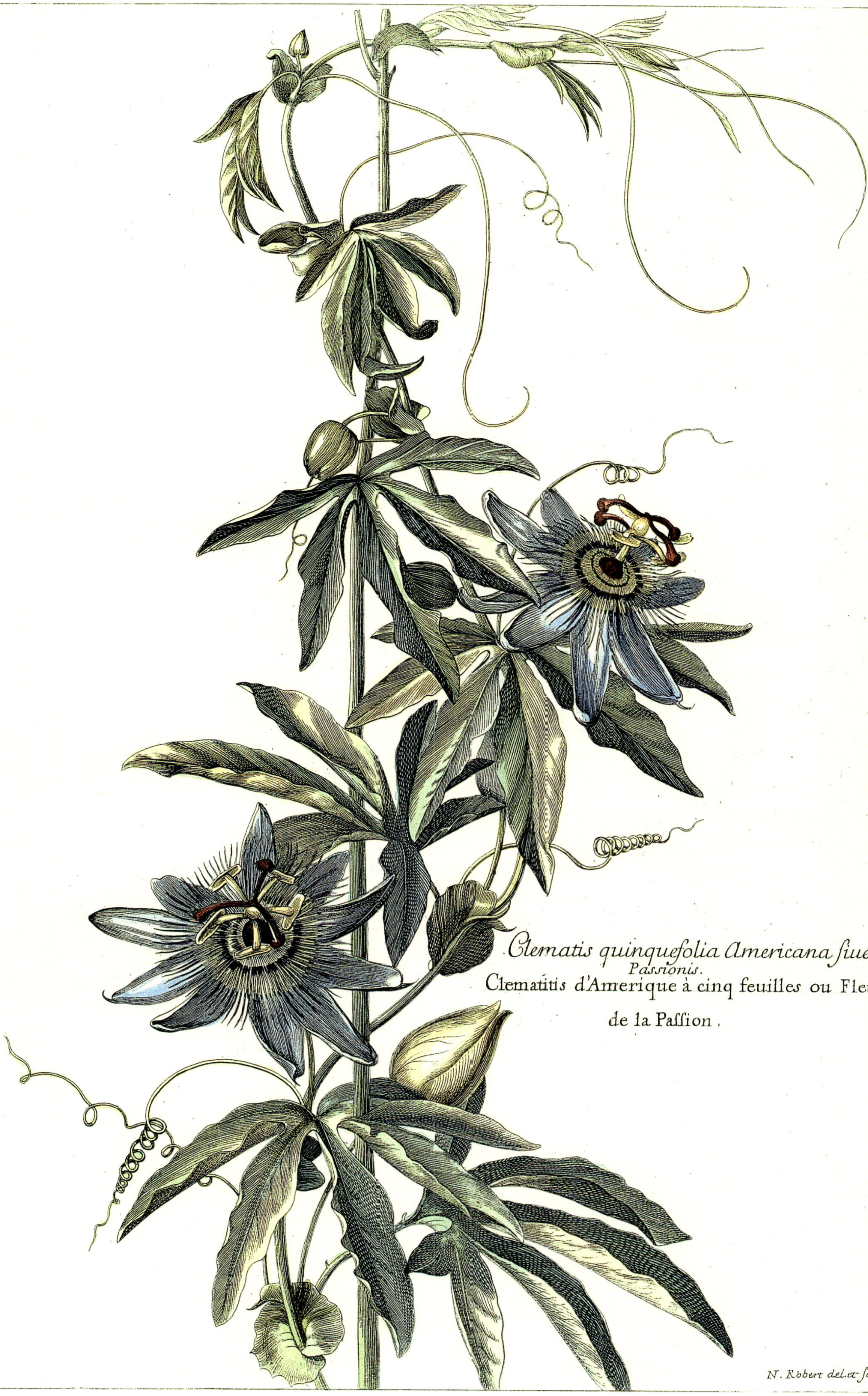

Clematis quinquefolia Americana siue Flos-Passionis.
Clematitis d'Amerique à cinq feuilles ou Fleur de la Passion.

N. Robert del. et sculp.

Passiflora incarnata L.

PASSIFLORACEAE

Passiflora incarnata L., the medicinal passionflower, is native to the southern United States (Virginia, Texas and Florida). It is concentrated mainly in the eastern regions, from northern Florida right up to the border with Ontario.

The plant was brought to Europe in 1608 by Captain John Smith, the heroic figure who featured in the animated film *Pocahontas*. He speaks of its pleasing, lemon-flavoured fruit, known in the New World as *maypop*.

Passion fruit soon became very popular indeed. Of the species with edible fruit, *P. edulis* and *P. ligularis* are the most sought after. Other species cultivated include *P. actinia*, *P. vitifolia*, *P. foetida*, *P. tarminiana*, *P. manicata*, *P. laurifolia* and *P. quadrangularis*, the fruit of which must be eaten when very ripe. The fruits of many other species are poisonous, though some only until they are fully ripe, for example *P. adenopoda* and *P. gibertii*. A number of cases of poisoning have been recorded, some of them fatal.

The first medicinal passionflower was planted in Paris in 1612, and then in Rome in 1619.

The wild plant is dried for medicinal use and only the stems are used. American doctors have studied the effects of the plant and used it in neurological applications. It contains flavones such as vitexol, orientol and saponarol and acts as an anti-spasmodic and sedative on the central nervous system. It is an effective anxiolitic, in homeopathy too, according to the celebrated Dr Leclerc who prescribed it for anxiety in the First World War. Recently, the passionflower has produced some interesting results in the treatment of opiate addiction.

Passiflora incarnata L. Engraving : Nicolas Robert
RMN, Louvre Chalcography Dept, Paris (modern print).

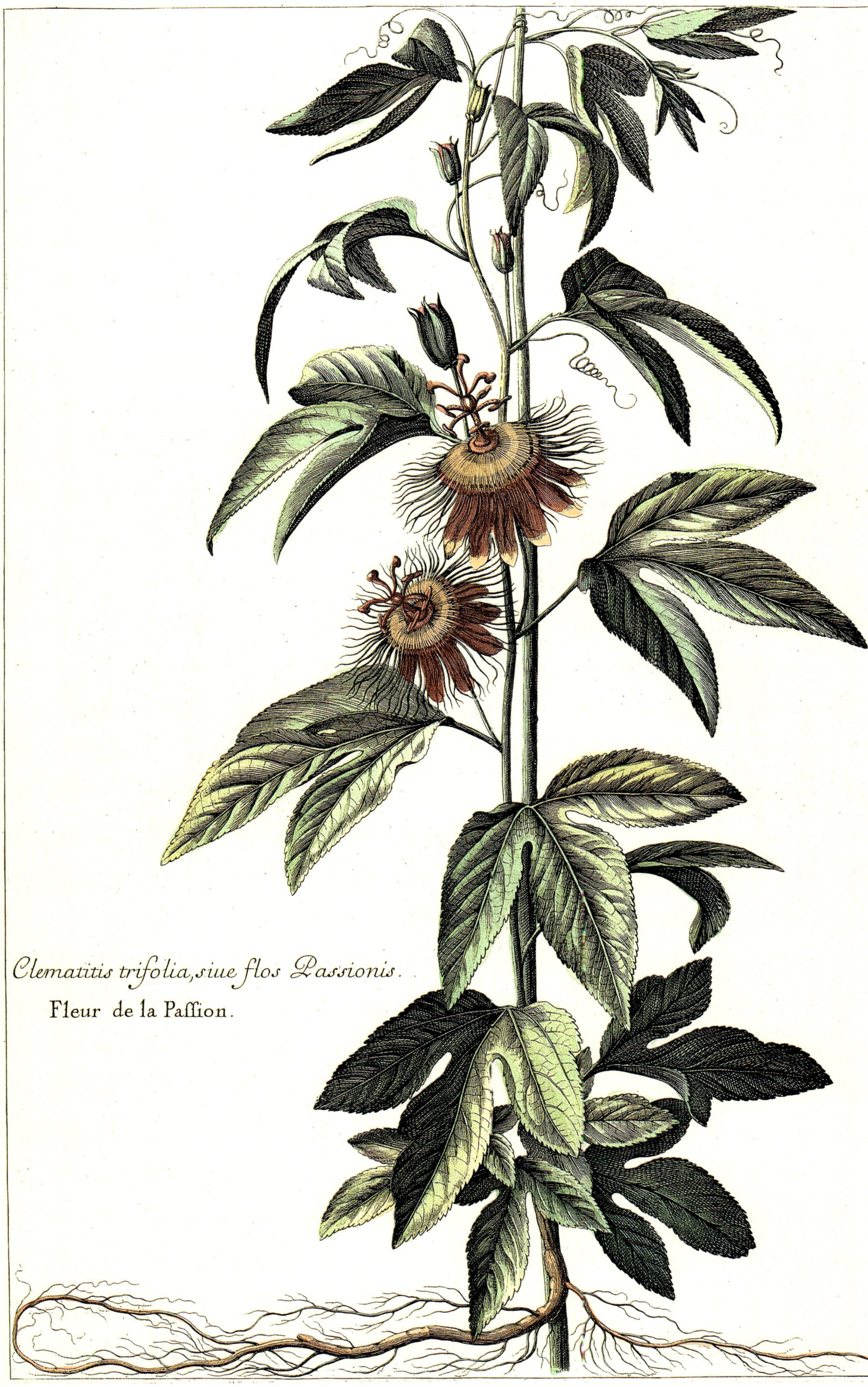

Clematitis trifolia, siue flos Passionis.

Fleur de la Paſſion.

N. Robert del. et ſculp.

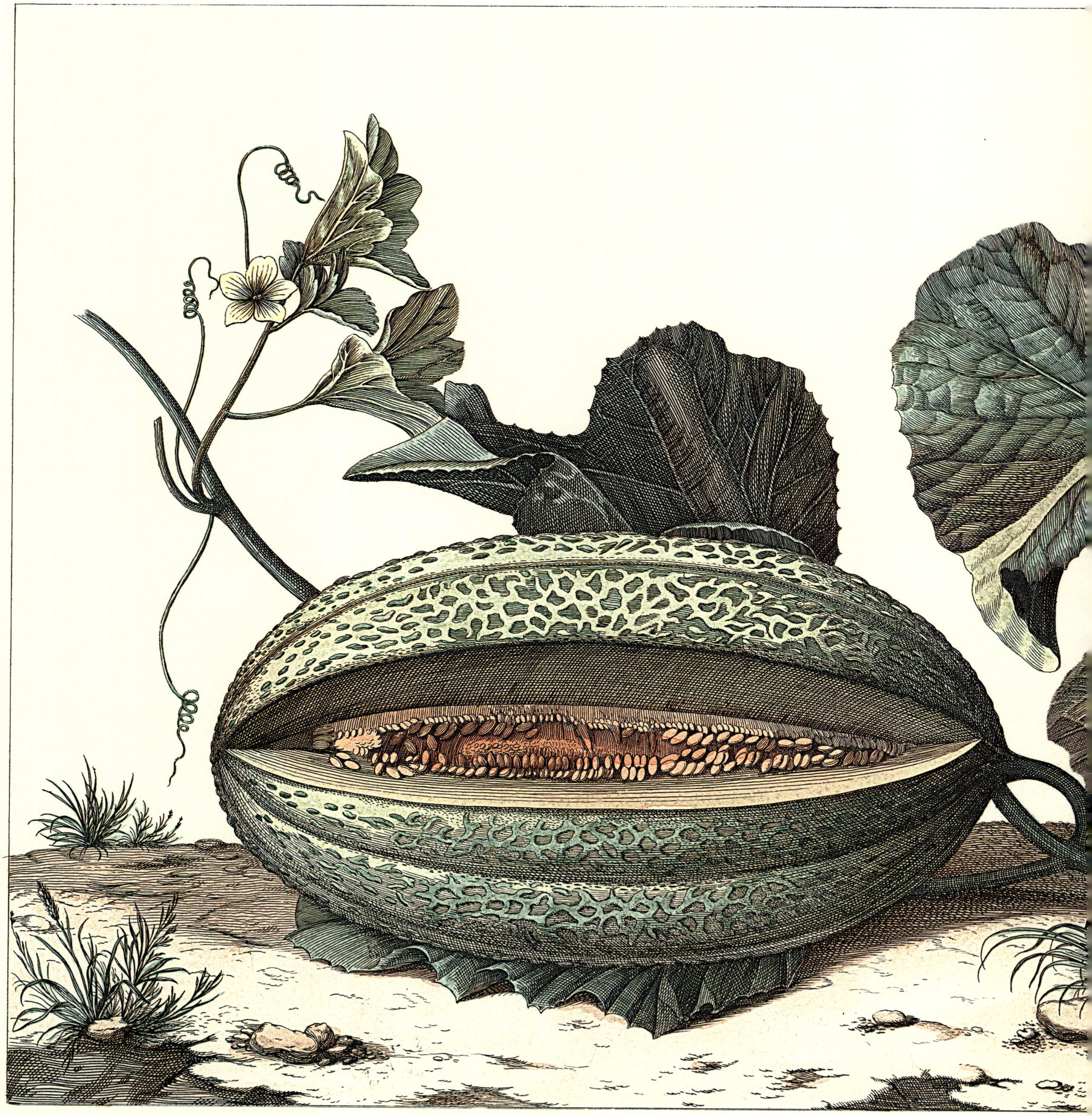

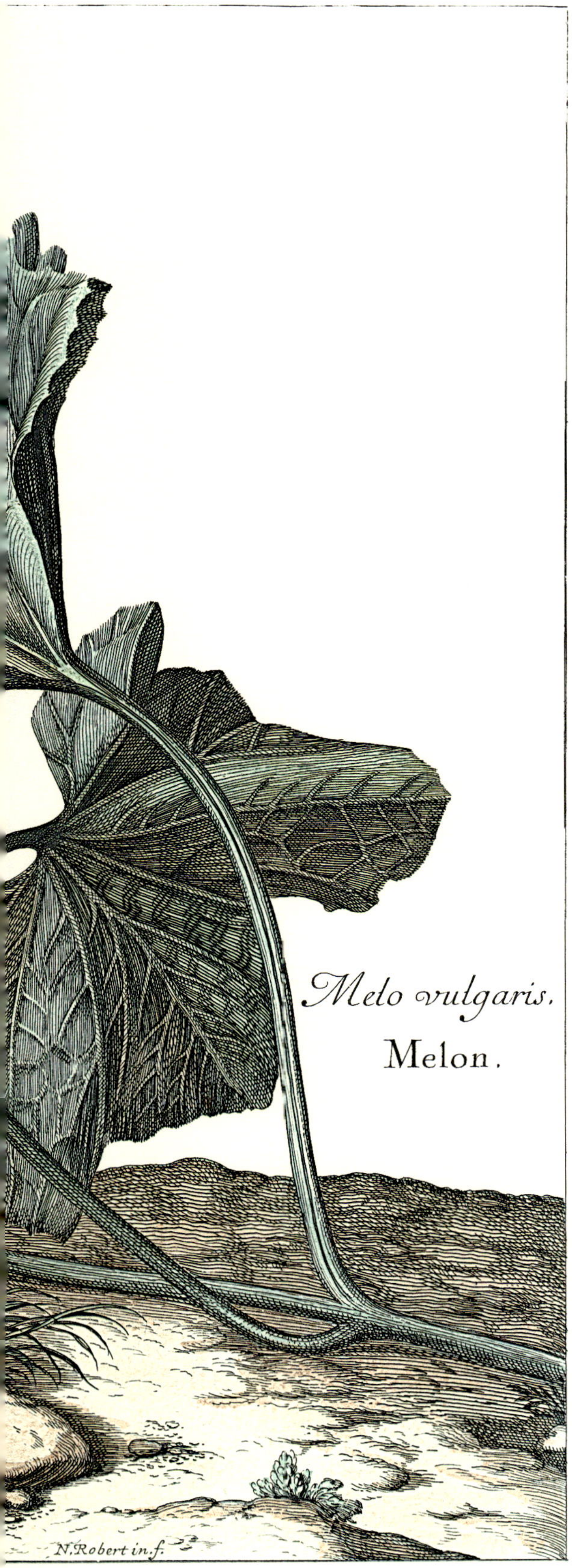

Cucumis melo L. subsp. melo

CUCURBITACEAE

No-one is certain where the melon originated – some believe it was in Iran, others India or southern Africa. The word 'melon' first appears in France in 1256 and comes from the Greek *melôpepôn*, from *melo* (fruit, apple) – from which the word *Malus*, the Latin for apple tree is derived – and from *pepôn*, which means 'cooked by the sun, ripe'.

In ancient times, the melon was cultivated in Asia and around the Mediterranean basin. Charlemagne's *Capitulare de villis vel curtis imperialibus* mentions four types of cultivated cucurbita: *cucumer* (the cucumber or gherkin); *pepon* (melon), *cucurbita* (the bottle gourd: *Lagenaria vulgaris* Ser.) and *coloquentida* (the bitter apple *Citrullus colocynthis* Schrad). In France, the name *pepon* had become *pompon* by the fifteenth century.

Here is what Daléchamps has to say on the subject in 1615 (book V. chap. XXX: *Des pompons & melons*): 'As for melons, there are also several kinds, some of which are large, others the size and form of a quince, sometimes with white flesh, sometimes flesh that is yellow or pale, sometimes soft. Most highly prized are those with firm, white flesh & so sweet that it seems as if they were preserved in sugar and these are also known as *sucrins* and *sucrez*. Furthermore, some have their seeds interspersed and firmly attached to the flesh, so much so that it is necessary to remove the seeds one by one with the tip of a knife; others have their seeds in a moist pith and fall out when the melon is shaken, separating easily from the flesh ... One can tell a good one because it smells good and is very heavy and has a large and very bitter stalk – a truly bitter stalk being an infallible sign of a good melon. They like a good, rich soil, in full sun and plenty of room to stretch out in comfort.'

The melon commonly eaten is a subspecies of *Cucumis melo* L. Selection has produced three types of melon that are eaten as fruit: the cantaloupe melon (*cantalupensis*); the winter melon (*inodorus*); and the netted melon (*reticulatus*). There are melons which are harvested before they are fully ripe and eaten as vegetables, but this is not common in France.

Melons are enjoyed all over the world. And although some people believe that they are a speciality of southern France, in fact France ranks only thirteenth in the world in terms of melon tonnage, lagging a long way behind China, Turkey, the United States and Spain.

However, the melon has experienced some curious declines in popularity in its history. In 1868, Cazin in his *Traité pratique des plantes médicinales et indigènes*, recounts how a certain doctor considered that the accidents caused by this fruit were so frequent and so serious that they made his fortune. He had the following inscribed in gold lettering on the magnificent house he had just had built: 'It is thanks to the cucumber and the melon that I have built this house.'

Cazin also mentions the medicinal use of the so-called 'greater cold seeds' for lung ailments and the calming properties of the fruit when applied to burns, swellings and other contusions. He also adds: 'According to Lange, a suppository made with a piece of melon or cucumber has been known to stop excessive haemorrhoidal bleeding.'

Cucumis melo L. subsp. *melo*. Engraving: Nicolas Robert
RMN, Louvre Chalcography Dept, Paris (modern print).

Alchemilla vulgaris L. (group)

ROSACEAE

The name alchemilla has a certain magical resonance. It is associated with the alchemists' quest for the philosopher's stone – the quest to turn metal into gold, the spiritual quest to find what is most noble in oneself – the absolute!

Alchemists would go out to collect the water held among the leaves of this plant, water they believed to be pure and celestial and endowed with marvellous properties. This was particularly true of the water collected on St John's Eve, the night before the summer solstice.

The word 'alchemy' comes from the Arabic *al kimiyâ*, which referred to the philosopher's stone and is derived from the Arabic root *kama* (to keep secret). Some believe it is Greek in origin, from *khêmia* (black magic), or Hebrew, from *semes* (sun). Caught then between light and darkness!

The Arab alchemists of the tenth century called the celebrated stone *al iksïr*, which became 'elixir' under the influence of the Greek word *xerion*, a dry medicinal powder (xeric), that is to say the purest extract of a substance. This word came to mean a drink with supposedly magical properties – for example the elixir of love or the elixir of long life – or a plant-based alcoholic liquor with medicinal properties – paregoric elixir, the Grande Chartreuse elixir, etc.

Alchemilla is rarely used today. However, it contains flavonoids, such as procyanidol, which improve capillary resistance, and tannins which are very useful in more prosaic disorders such as diarrhoea.

Alchemilla vulgaris L. (group). Engraving : Louis de Chastillon
RMN, Louvre Chalcography Dept, Paris (modern print).

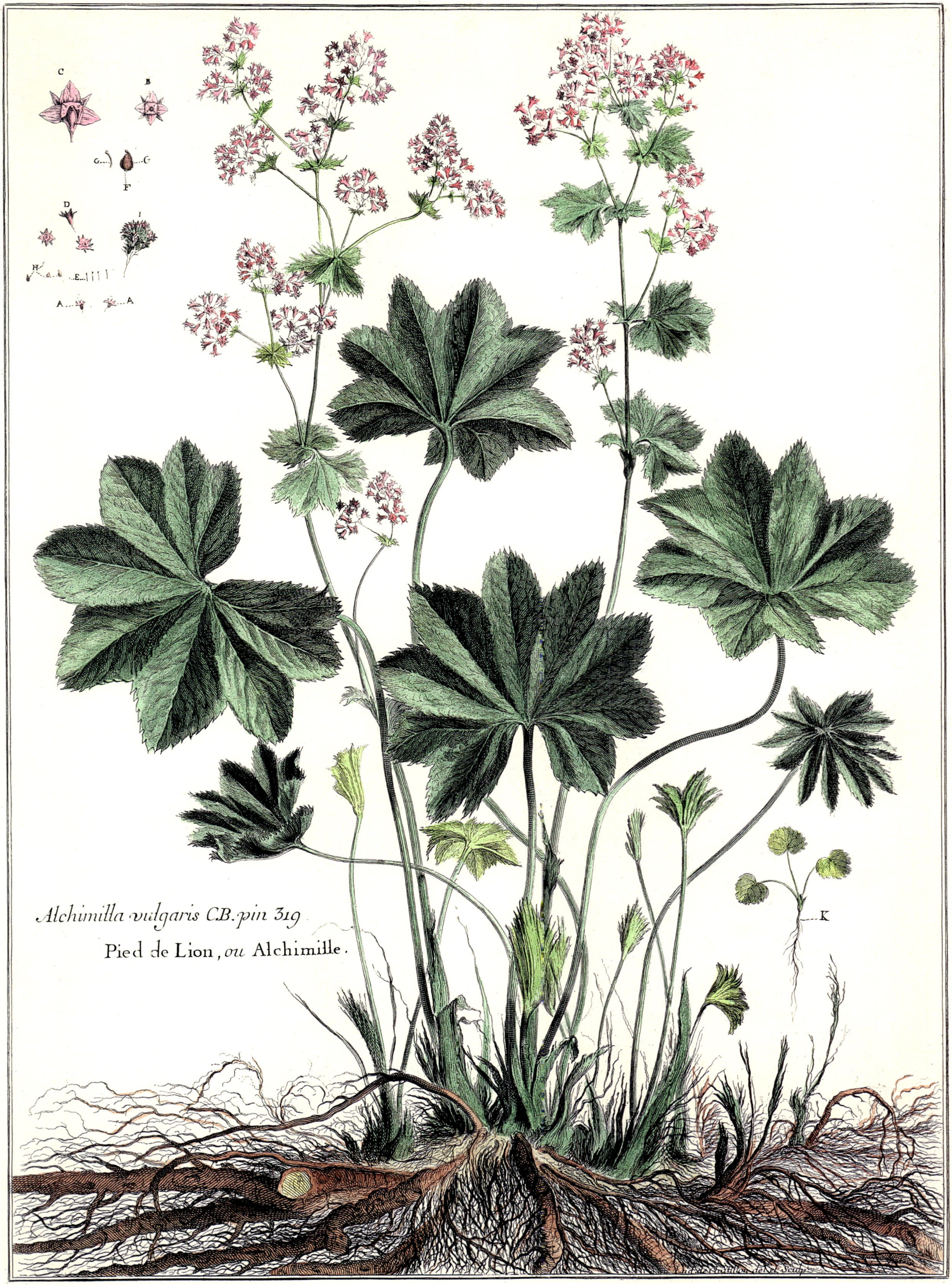

Alchimilla vulgaris C.B. pin 319

Pied de Lion, *ou* Alchimille.

Alchemilla vulgaris L. (group)

ROSACEAE

The alchemillas are divided into three groups: the *Alchemilla pentaphyllea* group which includes a unique species with deeply serrated leaf lobes; the *A. alpina* group and its two sub-groups *saxatiles* and *hoppeanea* in which the leaf lobes are divided down to the base or very nearly; and the *A. vulgaris* group which has shallow-lobed entire leaves. This last group which, with at least fifty or so species is the largest, includes two sub-groups, one containing species with pubescent leaves (the series *pubescentes*, *vulgares heterpodae*, *vulgares hirsutae* and *vulgares subglabrae*), the other with glabrous leaves (the series *calycinae*, *splendentes* and *vulgares glabrae*). *Alchemilla* is a complex botanical genus into which few botanists choose to venture.

The alchemilla has always attracted interest, both for its beauty and for its ability to collect dew in small, translucent, silvery droplets in the hollow of its leaves – this was the liquid that fired the imagination of alchemists and condemned them in their quest. These droplets of dew are not to be confused with the small beads of liquid that form around the rim of the leaves: these are exuded by the aquiferous stomata or hydrathodes – tiny, microscopic openings that enable the plant to discharge excess water. This phenomenon, known as guttation, is found in other species such as cabbage, ivy, tomato and nasturtium. The liquid emitted contains mineral salts (calcium in particular) and various compounds (organic, glucidic and nitrogenous) in small but variable quantities, depending on the plant.

In the Baltic countries, alchemilla, which is rich in tannins, was traditionally used as a yellow fabric dye. It is these gallic, ellagic and catechic tannins that give the plant its astringent and healing properties.

Daléchamps (1615) describes some of the uses to which alchemilla was put but which have now disappeared: 'Its leaves and root are highly astringent and desiccative ... and German surgeons make great use of them and put them in their vulnerary beverages and potions. They also use them in decoctions for cleansing wounds and even apply dressings soaked in such decoctions to these wounds; and also to breasts that are too soft and flaccid, to harden them and make them firm; for which they are very useful, especially if one adds hypocistis, horsetail, alum and dried roses. The powder of this dried herb taken with the decoction or water distilled with this herb will help to cure breakages and aid digestion in young children, being a good strengthener. A spoonful of this same powder taken in wine or in broth for fifteen to twenty days is most proper for women who are unable to conceive because their womb is too moist and are therefore unable to retain the man's seed. Water distilled from this herb and taken either as a beverage or introduced into a woman's womb is marvellously proper for easing the white flux; in fact, if continued this remedy will make a woman like a virgin, even though she may not be, so strongly will these parts be tightened. What is even better and takes less time, is if the vapour from a decoction of this herb is introduced internally.'

Alchemilla vulgaris L. (group). Engraving : Nicolas Robert
RMN, Louvre Chalcography Dept, Paris (modern print).

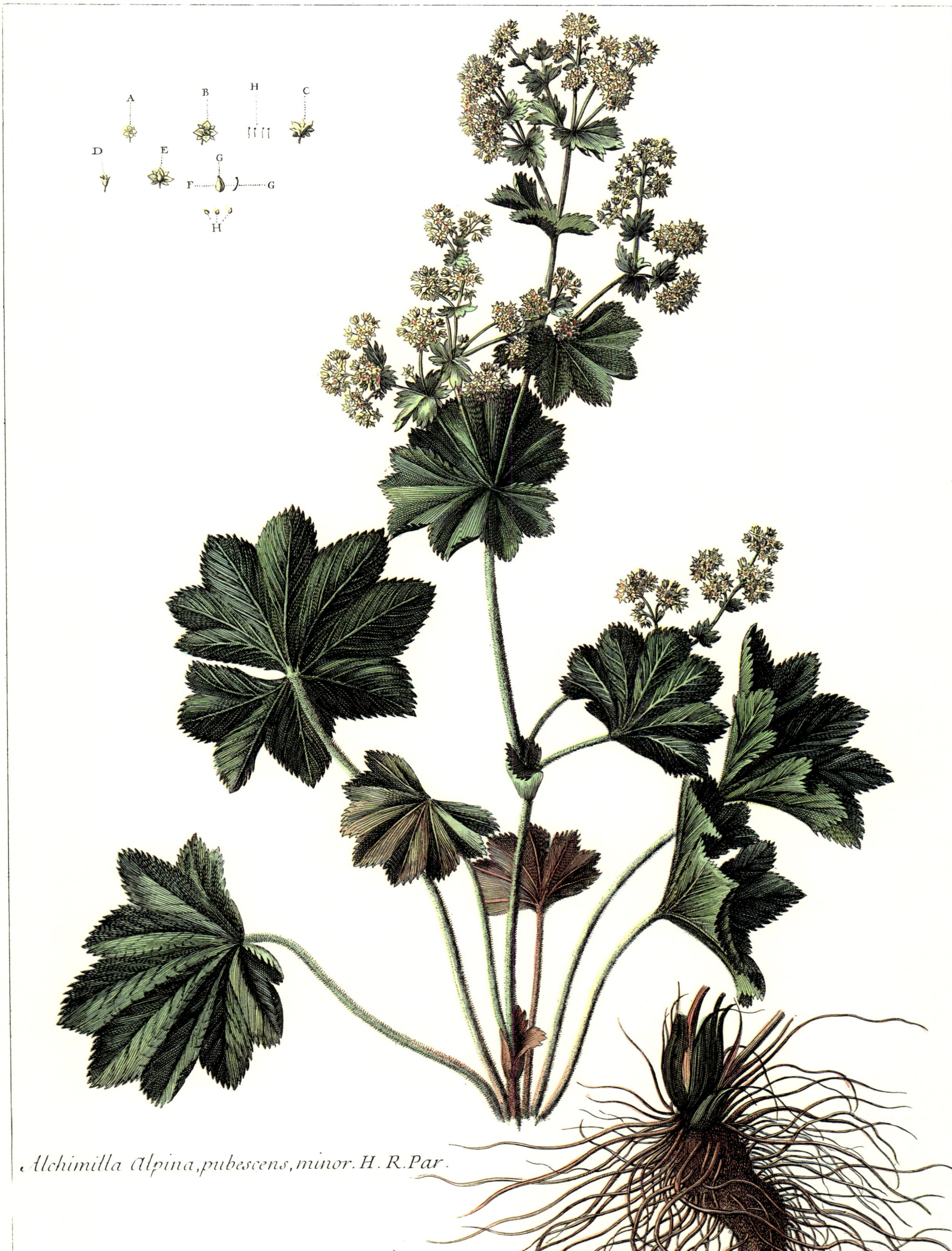

Alchimilla Alpina, pubescens, minor. H. R. Par.

N. Robert del. et sculp.

Physocarpus opulifolius (L.) Maxim.

ROSACEAE

It took some time before a name for this plant could be agreed. A species of spirea with guelder-rose leaves, it was variously known as *Physocarpus stellatus* (Rhydb.) Redher, *Spirea opulifolius* L., *Neillia opulifolia* (L.) Brewer & S. Watson. In 1753, Linnaeus settled on *Spirea*.

The plant originates in North America, in the eastern United States (Virginia, Carolina, Dakota, Oklahoma, Wisconsin, etc.) extending as far north as Quebec, where it is found in the cities of Quebec, Montreal and beyond. It was introduced into Europe in 1687.

In Quebec it is known in French as *bois aux sept écorces* (sevenbark), but English speakers call it 'ninebark': common ninebark, eastern ninebark or smooth ninebark. The highly decorative bark of this shrub peels away in strips to reveal brightly coloured layers below. Sometimes the bark splits vertically and forms scar tissue. The youngest branches are a reddish brown.

The green leaves of spring turn an attractive bronze in autumn. This hardy, elegant shrub grows spontaneously near rivers and is planted in gardens to make hedges and shrubberies. Its handsome corymbs of small, white flowers are popular with bees and butterflies.

There are thirteen species of *Physocarpus* in North America and one species in south-east Asia. Many horticultural varieties are sold across the world. Examples include 'Dart's Gold', which has golden yellow leaves in autumn; 'Diabolo', which is a handsome, intense purple, almost black; and 'Luteus', an old variety dating from 1864 which is a fine, yellowy bronze.

It was taken medicinally by the native peoples who used the leaves and the second layer of bark as a purgative and emetic. Its antibacterial properties made it useful as a poultice for skin disorders and scabies, for female hygiene and in cases of gonorrhoea.

The name *Physocarpus* comes from the Greek *phusa* (meaning vesicle or bubble), an allusion to the shape of the fruit; and from *karpos* (fruit) – fruit also referring to the harvests of the earth. It is also associated with the French *cueillir* (to gather, to pick), the High German word *herbis* (autumn, hence the month of gathering) and *harvest*, the English term for these various activities.

A vast number of botanical terms relating to fruit derive from the word 'karpos', for example pericarp and endocarp. The well-known saying *Carpe diem (quam minimum credula postero)*, taken from a poem by Horace, is often understood to be encouraging hedonism in its broadest sense: 'Seize the day.' However, it should in fact be translated as 'harvest the day', that is to say 'savour this moment', in a noble epicurean sense.

Physocarpus opulifolius (L.) Maxim. Engraving : Nicolas Robert
RMN, Louvre Chalcography Dept, Paris (modern print).

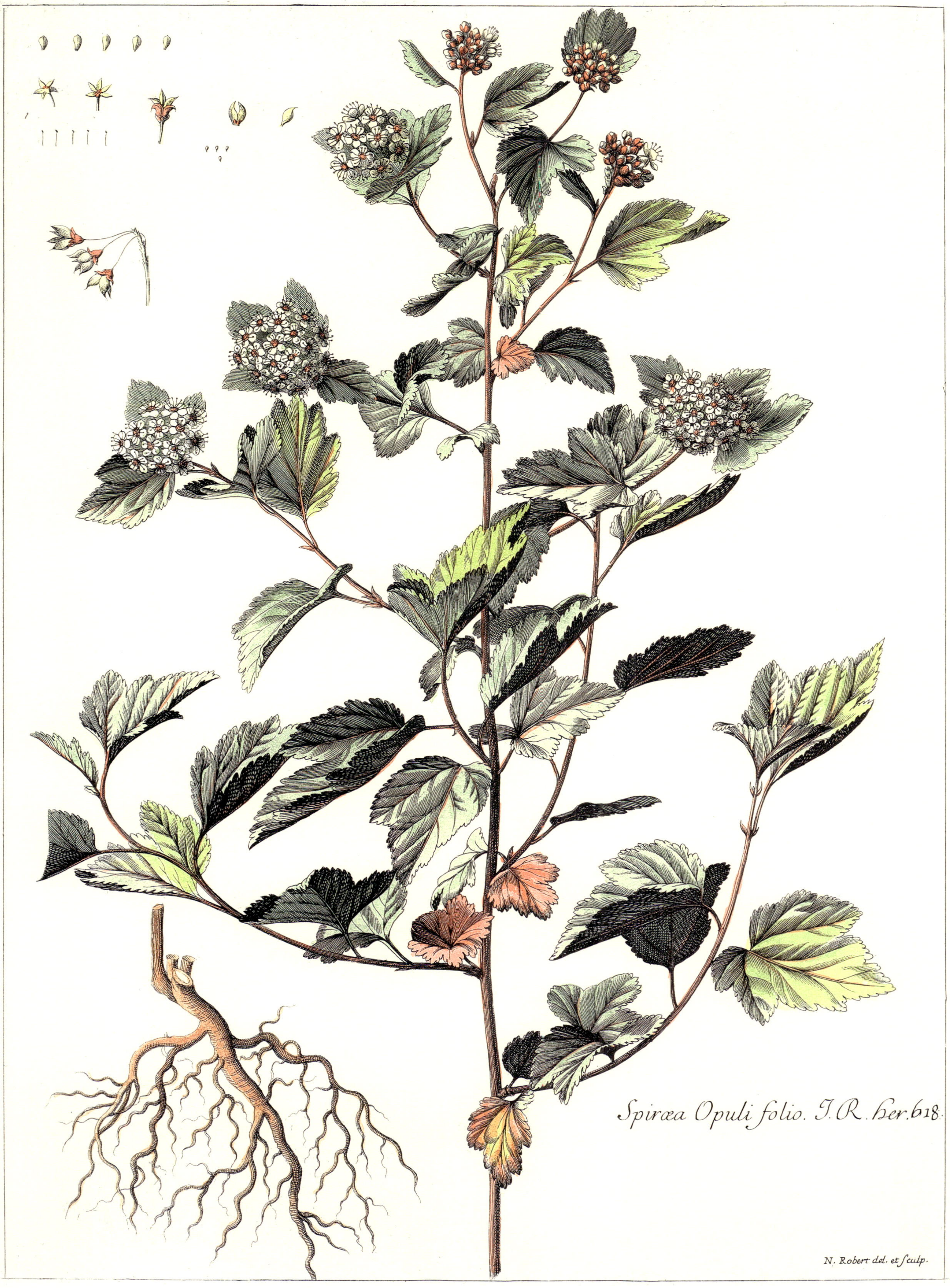

Spiræa Opuli folio. I. R. herb. 618.

N. Robert del. et sculp.

Prunus mume Sieb. & Zucc.

ROSACEAE

It would seem that almost every flowering tree that came from the Orient was classified as a 'Chinese hawthorn'! However, the only thing that the plant shown here has in common with the hawthorn is that it belongs to the family *Rosaceae*. The fact that it blossoms before the leaves appear would suggest a type of *Prunus*, which are highly prized in eastern Asia.

It is most probably a *Prunus mume*, a tree that plays an important role in Japanese culture. The Latin species term *mume* is inspired by the phonetic transcription of the name by which the tree is known in China and Japan. It is also known as the 'Japanese apricot'. In fact, the *Prunus* genus includes a very large number of trees and shrubs that are familiar to us: *Prunus amygdalus* Stokes is the almond tree; *Prunus armeniaca* L. is the apricot tree; *Prunus avium* Moench, is the wild cherry; *Prunus cerasus* L. is the sour or dwarf cherry; *Prunus domestica* L. is the plum tree; *Prunus laurocerasus* L. is the cherry laurel; *Prunus persica* (L.) Batsch is the peach tree; and *Prunus spinosa* L. is the blackthorn (sloe). It is therefore hardly surprising that this *Prunus* should have been labelled 'apricot' while its fruit is considered a plum.

It is a magnificent tree, growing to a height of 7 to 10 metres (23 to 33 ft), with a spreading crown when mature, covered in beautiful, delicate pink blossom in spring. Cultivated, pruned specimens recall our peach trees when they are in bloom, and the blossom-covered branches of peach trees are much in demand for use in *ikebanas* (Japanese-style flower arrangements) and for Japanese-inspired art photography. The leaves of *Prunus mume* are oval to round, terminating in long, fine points. Native to China and Korea, it was introduced into Japan in 1844. Many horticultural varieties now exist, each as attractive as the other.

Prunus mume may mean very little to most of us, except perhaps to vegetarians and people who frequent healthfood shops. In fact, the fruit of this tree, known as *ume*, is used to make the celebrated pickled plums known as *umeboshi* or *umebosi*, and which are sold as whole fruits garnished with a handful of leaves or as a paste.

In Japan, *umeboshi* are very popular. The fresh fruit, which is yellowish green, takes on an orangey red colour when salted: once washed and dried they are placed in a dish and doused with brandy before being sprinkled with 200 grams of salt per kilo (3 oz per lb) of fruit, along with salted, lactofermented *shiso* leaves. Once thoroughly mixed together, they are covered with a close-fitting lid and a weight placed on top to ensure that they remain completely submerged in the liquid. The salting process takes about four to five weeks and the plums become very *yang*.

Shiso is a small lamiaceae resembling a Coleus, with the botanical name *Perilla frutescens* (L.) Britton. Its leaves bring out the brownish red colour of the fruit and, most importantly, contain citral, limonene, linalol, estragol, etc. – substances also contained in more familiar aromatic labiates such as thyme, balm and basil. Linnaeus (1707–78) in fact called the plant *Ocimum frutescens* L., relating it to basil (*Ocimum*). The quality and flavour of the *umeboshi* depends on the *shiso*. Some varieties are more sought after than others, particularly the *crispa* varieties of which there are many forms: the true Chinese *shiso*; the more commonly eaten *shiso*; the *Ata-chirimen-jiso*; the *Chirimen-katame-jiso*, and a few others.

One traditional dish is made by arranging a circle of *umeboshi* on a bed of white rice in a symbolic representation of the Japanese flag.

Prunus mume Sieb. & Zucc. Engraving : Louis de Chastillon
RMN, Louvre Chalcography Dept, Paris (modern print).

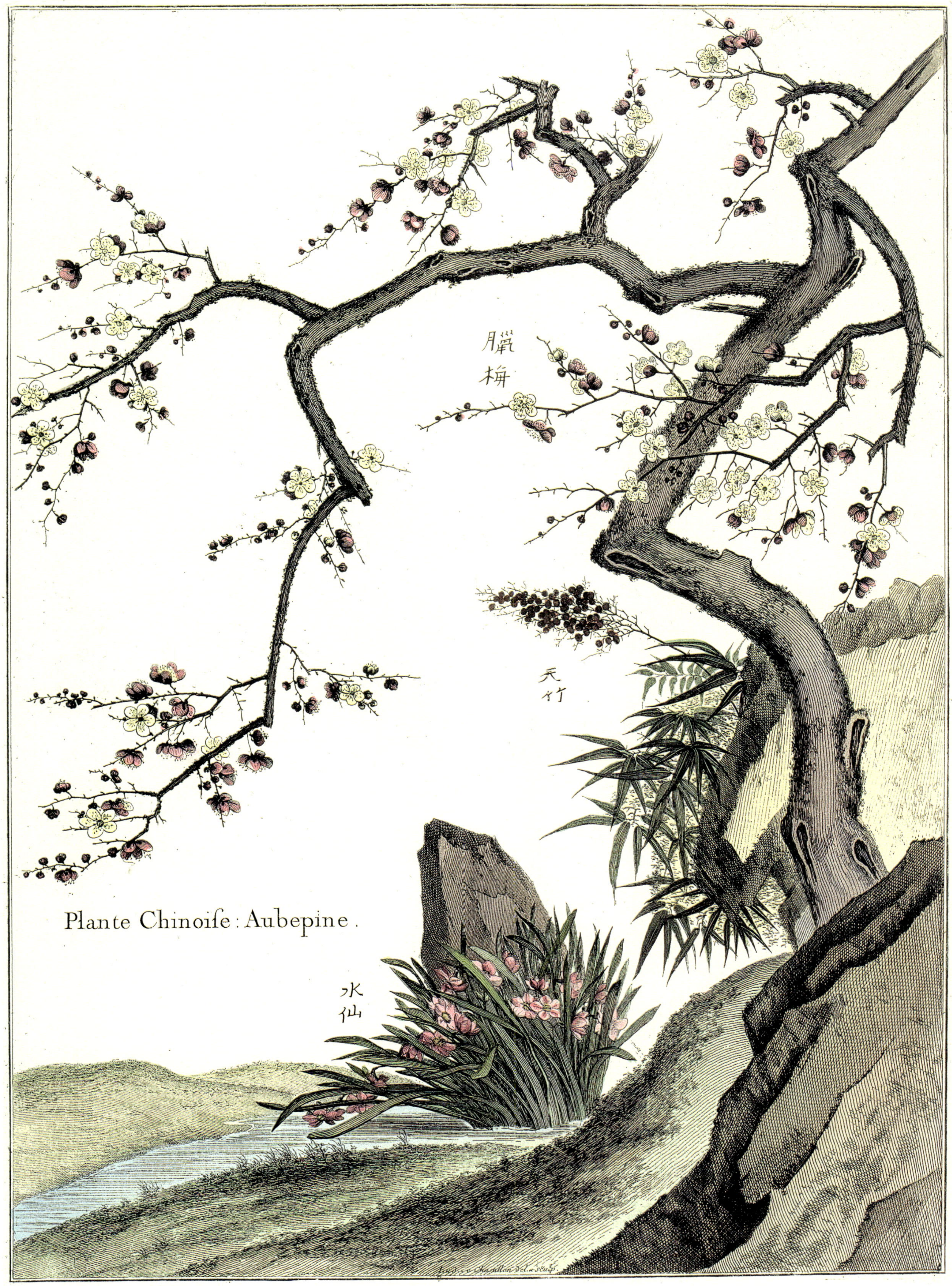

Plante Chinoiſe : Aubepine .

Malus spectabilis (Aiton) Borkhausen

ROSACEAE

It is fortunate that when Louis de Chastillon made this engraving in 1676 he added the name of the plant in Chinese to the right of the trunk because, on its own, the term 'Chinese plant: Hawthorn' would be misleading. This specimen does not appear to correspond to any known hawthorn. It might seem similar to *Crataegus wattiana* Hemsl. & Lace because of its large blossoms, but its leaves are divided into several lobes. Other hawthorns have corymbs of much smaller flowers, as do European hawthorns. To call it 'Chinese plant: remarkable apple tree' would seem more appropriate.

In fact, this illustration has more in common with an apple tree and this is confirmed by the Chinese name, the phonetic transcription of which – *Hai tang* (or *Hai tang hua*) – refers to the Chinese flowering crab apple or, to be more precise, the *ten thoa hoc* or *Malus spectabilis*, one of the most remarkable of the flowering apples. When the tree is mature, it is transformed in spring into a vast bouquet of large pink blossoms and truly lives up to its Latin description of 'spectacular'.

The blossom is magnificent – bright red buds opening into pinkish white to pure pink flowers reaching as much 5 centimetres (2 in) in diameter, with edible fruit from yellowish to pure yellow. The fruit, which is a little too acid when fresh, is best left to ripen or can be cooked. The leaves are narrower than those of our apple trees and are firm and glossy in appearance, elliptical in shape, with a short point.

Flore de Chine (1995) describes this plant as originating in the regions of Jangsei, Heibei, Liaoning, Quinghai, Shandong, Shanxi,Yunnan and Zeijang where it grows happily at altitudes of between 500 and 2,000 metres (1,650 and 6,500 ft). The tree, which can grow to 8 to 10 metres (26 to 33 ft) in height, was introduced into Europe in 1780. At that period, when apple trees and pear trees were confused as one genus, it was known as *Pyrus spectabilis* Aiton (Botanic Gardens, Kew, London, 1789).

One can easily imagine the impression it must have made and the desire it inspired to produce ever more floriferous cultivars. The two best-known varieties are *riversis* G. Kichner, with double pink flowers, and *albipena* Schelle, with double white flowers. It has been crossed with *Malus pumila* Miller, among others, to produce the hybrid *Malus magdeburgensis* which has pink flowers and small yellow to red fruit that are highly decorative in winter, and with *Malus baccata* (L.) Borkh., to obtain *Malus micromalus* Makino, with its fragrant white flowers.

The etymology of *Malus* is addressed on the page devoted to *Cucumis melo* (p. 51). As for the French word *pomme* (apple), it comes from the classical Latin word *pomum*, which refers to edible fruit in general.

Spectabilis is derived from the Latin *spectare* (to look at, contemplate), from the old European root *spek* (to observe), which has given us such words as inspect, respect, respite, aspect, species, spice, epitaph, spectre and spectator.

Malus spectabilis (Aiton) Borkhausen. Engraving : Louis de Chastillon
RMN, Louvre Chalcography Dept, Paris (modern print).

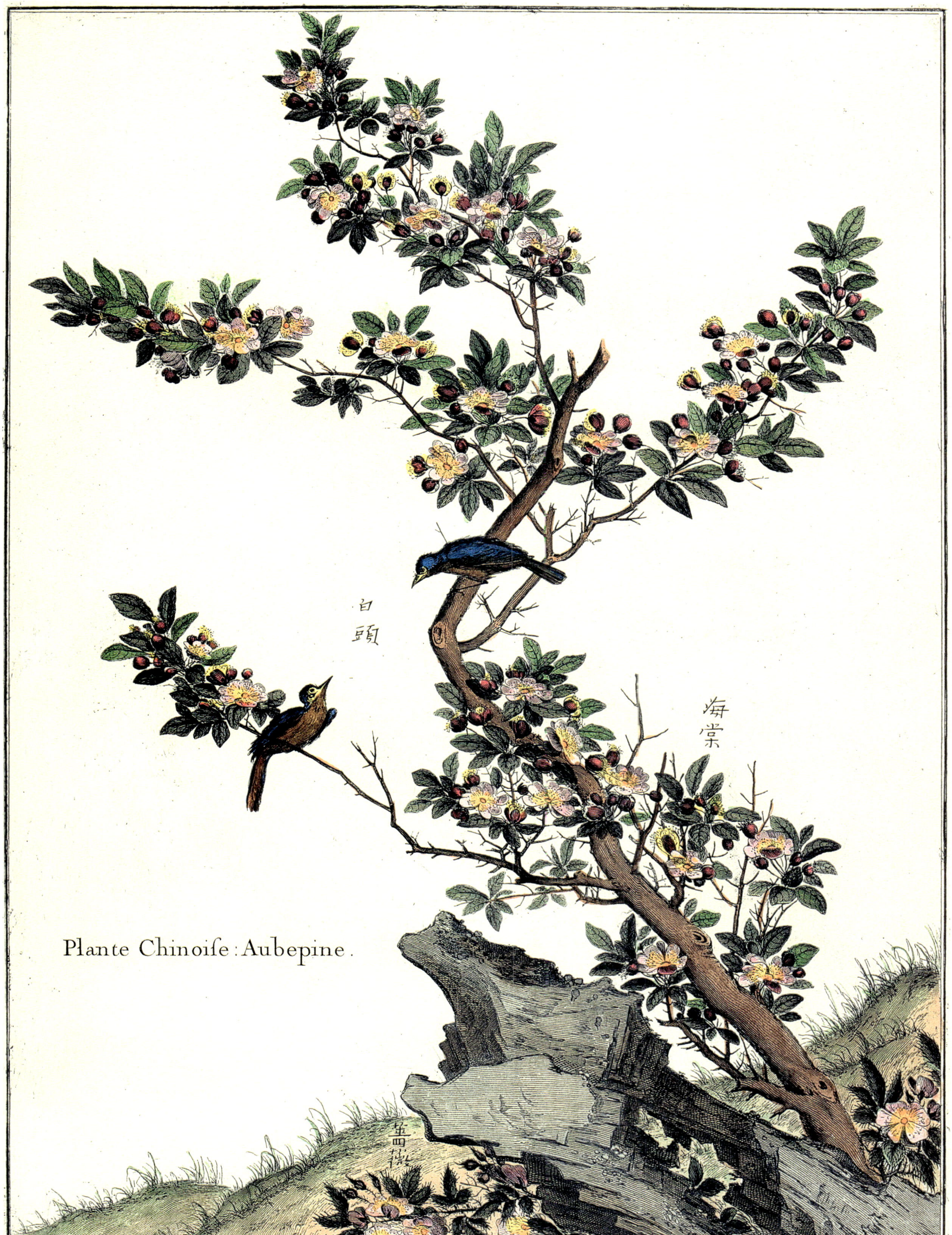

Plante Chinoise : Aubepine.

Astragalus monspessulanus L.

FABACEAE (EX-LEGUMINOSAE AND PAPILIONACEAE)

When we hear the word 'astragalus' we probably think first of the anklebone rather than the plant, unless of course we are part of the little world of botany! It is on this small but very important bone that our body essentially rests. Depending from which side one looks at it, it resembles the head of a dog or an ibex. Its name comes from the Greek *astragalos* ('vertebra' and 'heel bone') and is related to *osteon* (bone). Latin has added the figurative sense of 'moulding' by associating it with the knucklebones once used in the very popular game of that name and which came from the carpal bones of pigs and other animals. Certain botanists associated the plant with astragalus metaphorically, some because its seeds resemble knucklebones, others because the shape of the pod and the knotted stems are reminiscent of vertebrae. The least interesting astragalus (more commonly known as milk vetches) was known in French as the 'esparcette bâtarde'. In the past, the word 'esparcette' referred to the forage crop sainfoin and was derived from an old Provençal word meaning 'spaced', an allusion to its loose, open inflorescences.

Species of the genus *Astragalus* have elected to grow in difficult terrains – dry, arid, rocky areas in the south-east of France, scree and rock-covered ground in high mountain areas and coastal sands. All of them situations that require a certain capacity to adapt. The milk vetch is able to do this because of its substantial roots, its small leaves divided into folioles and, most importantly, its ability to draw nitrogen from the air thanks to nitrogen-fixing bacteria located in its root nodes.

As a member of the legume family, the milk vetch is characterised by long, narrow leaves, most often imparipennate with numerous entire leaflets. The flowers are grouped in axillary clusters, sometimes almost sessile, sometimes carried on a long peduncle. The flower's standard is longer than the wings, which are themselves longer than the keel, which is never pointed. One often sees peduncles lying flat against the soil, in a radiating fashion, the flowers and pods upright.

There are more than 1,350 species of vetch, distributed across the northern hemisphere, South America and Africa. In China, pieces of root of *Astragalus hoantchy* Franch. are sold as a tonic and diuretic under the name *Hoang k'i*. In France, some species were once used as a source of good, natural forage; and some, such as the vetch shown here and the liquorice vetch, are now cultivated as ornamentals in rockeries and gardens. In the past the vetch was used for its diuretic and sudorific properties to treat rheumatism and gout.

From the 1920s to the 1950s, a famous hair preparation known as 'gomina' and sold in the form of a powder to be mixed with water, was very much in vogue. All men, young and old alike, used it to slick down their hair, to make it smooth it and hold it in place. More recently there has been a return to this fashion among adolescents with the use of hair gel, but now, instead of smoothing their hair back in the style of old politicians, they make the most of its stiffening properties to produce outlandish hairstyles. The 'gomina' of old was based on gum tragacanth, an exudant that hardens on contact with the air and is found in the branches and trunks of the spiny shrub, *Astragalus gummifer* Labill., and other members of the genus.

Astragalus monspessulanus L. Engraving : Nicolas Robert
RMN, Louvre Chalcography Dept, Paris (modern print).

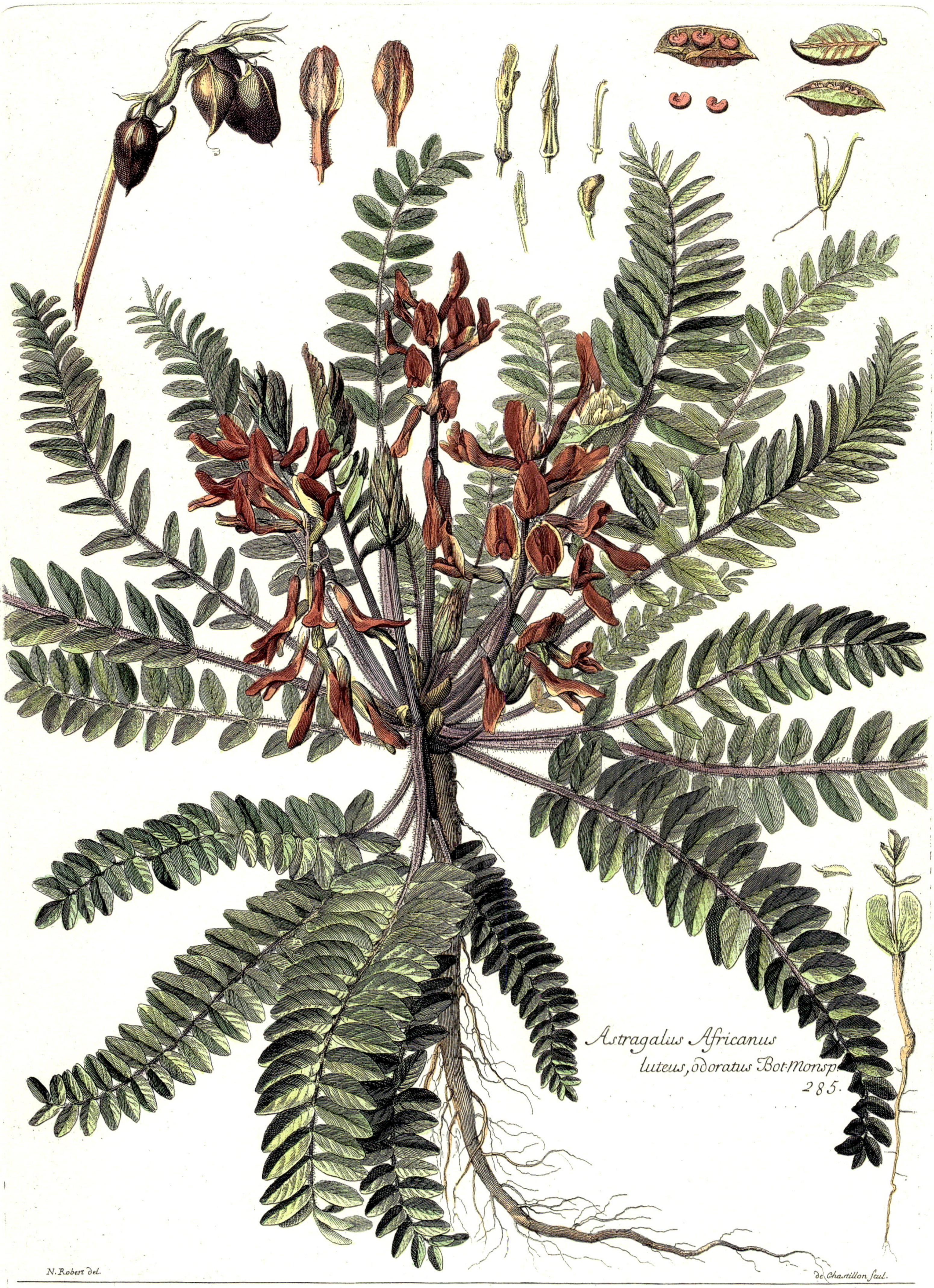
Astragalus Africanus
luteus, odoratus Bot. Monsp.
285.
N. Robert del.
de Chastillon sculp.

Medicago laciniata (L.) Miller et Trifolium subterraneum L.

FABACEAE (EX-LEGUMINOSAE AND PAPILIONACEAE)

Before they fruit, it is possible to confuse certain species of these two genera, but once they bear fruit all doubt is removed. With a few exceptions where the fruit are reniform (kidney-shaped) or falciform (sickle-shaped), the *Medicago* species, also called lucerne, have very distinctive rounded, corkscrew-shaped pods.

On the left of the plate, the *Trifolium echinatum, luteum, folijs laciniatis* is in fact *Medicago laciniata*, a species considered to be Saharan, or more precisely 'Saharosindh', the Sindh being one of the four provinces of Pakistan. It is an annual, growing from 10 to 30 centimetres (4 to 9 in) in height, with small, yellow flowers and leaves composed of three leaflets that are deeply dentate around the rim. The pods are extremely spiny, forming four to five quite slender whorls. Very common throughout the Sahara, the cutleaf medick is naturalised in a few places in the south of France, along the coastal Alpes-Maritimes and, more rarely, on the coast of Languedoc-Roussillon.

The name *Medicago* was given by Daléchamps in 1587 and confirmed for certain species by the French botanist Joseph Pitton de Tournefort (1656–1708). Daléchamps (1615) explains the origin of the word: 'In Latin, sainfoin is known as *Medica*, and in French as sainfoin and *foin de Bourgogne* [Burgundy hay] ... Pliny states that the name *Medica* is an adequate demonstration of its source. And indeed it did not exist in former times in Greece, and was brought there from Medea following the war between the Greeks and Darius, king of Persia.'

The *Trifolium gastonium* owes its name to the fact that it was discovered and described for the first time – a rarity in the history of plants – by a nobleman of royal blood: Gaston d'Orléans, brother of Louis XIII. Unfortunately for his descendants, this species name was removed and replaced with *subterraneum* because of a particular botanical feature: each small group of faded flowers and fruits becomes buried when the floral peduncle carrying them bends over and downwards. This group is composed of sterile flowers on the upper part and fertile flowers below. The fertile flowers obviously produce the fruit but these are covered by faded sterile flowers. Each bizarrely constituted group is then brought flat to the ground and when growing in sandy soil becomes partially buried, hence the French name *trèfle semeur* (sower clover) and the English term 'subterranean clover'. The seeds then germinate close to the base of the mother plant. This creeping clover with its white flowers streaked with pink is found in siliceous terrains, in meadows, on embankments, at the edges of fields and in fallow ground in most areas of France, but predominantly in the south, west and north-west.

Lucernes and clovers, along with other legumes, have a beneficial relationship with a genus of bacteria known as *Rhizobium*, which are able to fix atmospheric nitrogen. Accommodated in small root nodules, these bacteria supply their host with valuable nitrogen. Initially, a kind of complex molecular dialogue is established between the host plant and the *Rhizobium* to agree the terms of this symbiotic relationship and following this strange consultation process, nodes form on the roots to accommodate the bacteria.

Medicago laciniata (L.) Miller and *Trifolium subterraneum* L. Engraving : Abraham Bosse
National Museum of Natural History, Paris.

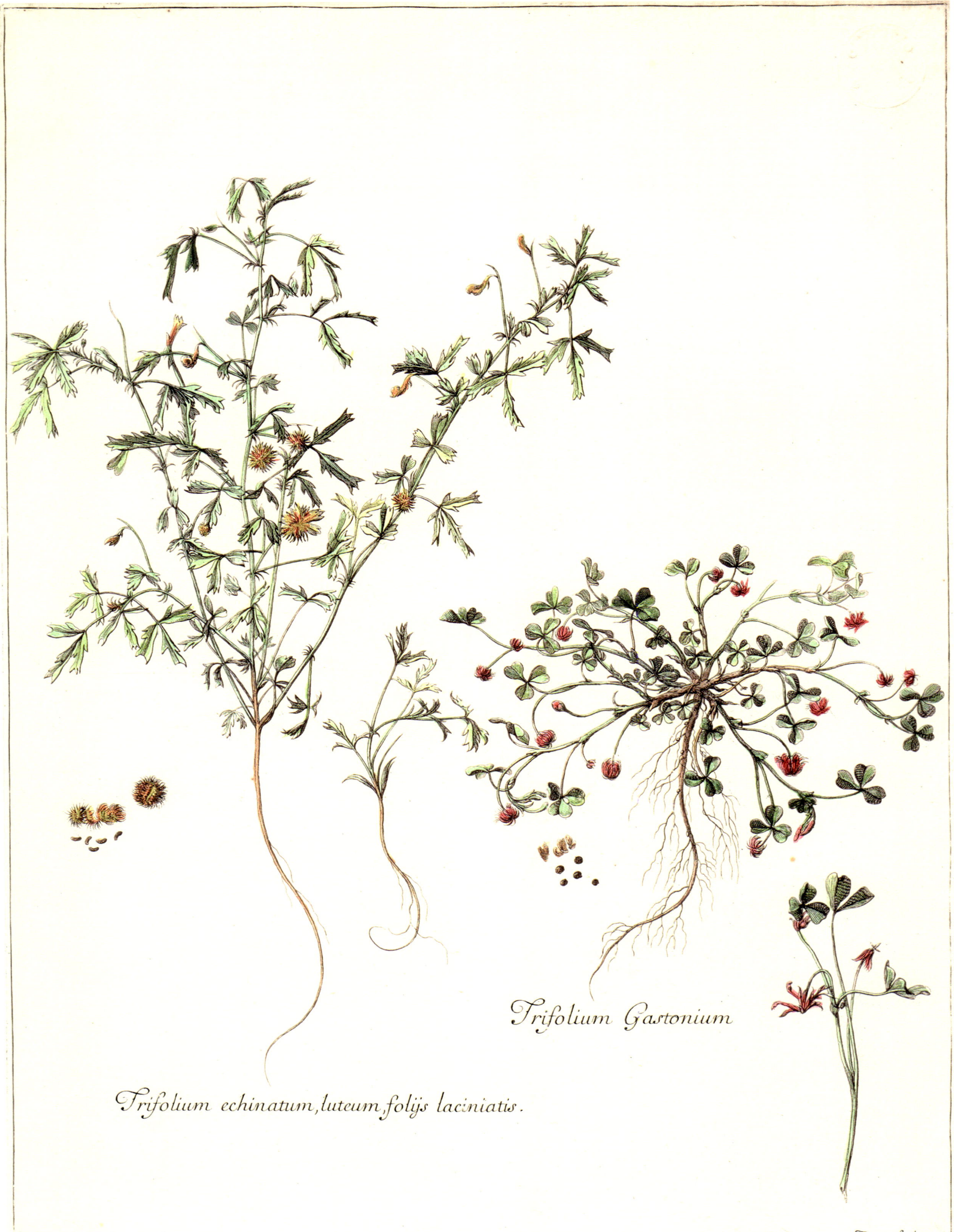
Trifolium Gastonium
Trifolium echinatum, luteum, folijs laciniatis.
ABosse sculp.

Ononis spinosa L.

FABACEAE (EX-LEGUMINOSAE AND PAPILIONACEAE)

Anonis, the former Latin name of this genus, is a reference to the fact that donkeys love eating this plant – from the Greek *onos* (ass) and *onemi* (delight). There are some sixty or so species growing in Europe, western Asia and North Africa, including almost twenty species in France, where it is commonly known as *bugrane*.

It goes relatively unnoticed by most people, yet the delicate, dark mauve lines against the pale lilac background of its petals give it a certain charm.

This small, spiny subshrub has a root system once reputed to bring an ox-drawn plough to a halt, hence its common name, restharrow. In the Middle Ages (1379) it was known in France as *bouveraude*, a combination of the Latin words *bos* (ox) and *retinere* (that which holds back). The French term *bugrane* comes from the Greek *boukranion* (ox head) which also refers to the ancient helmet of this shape.

This leguminous plant can adapt to living in very inhospitable terrain thanks to the presence of the nitrogen-fixing bacteria in its root nodes. The roots also harbour an essential oil containing transethanol, a substance that helps to eliminate sodium and chlorides via the urinary tract.

Here is a short etymological anecdote to conclude this plant portrait: the root *bos* (ox) is found in *boton* (herd animal), which gives the Greek word *boskein* (plant). Originally this term referred to plants used as grazing and fodder to feed livestock. It gradually developed into *botanê*, referring to plants in a more general sense, and eventually into the highly apposite term 'botany'.

Ononis spinosa L. Anonymous engraving
RMN, Louvre Chalcography Dept, Paris (modern print).

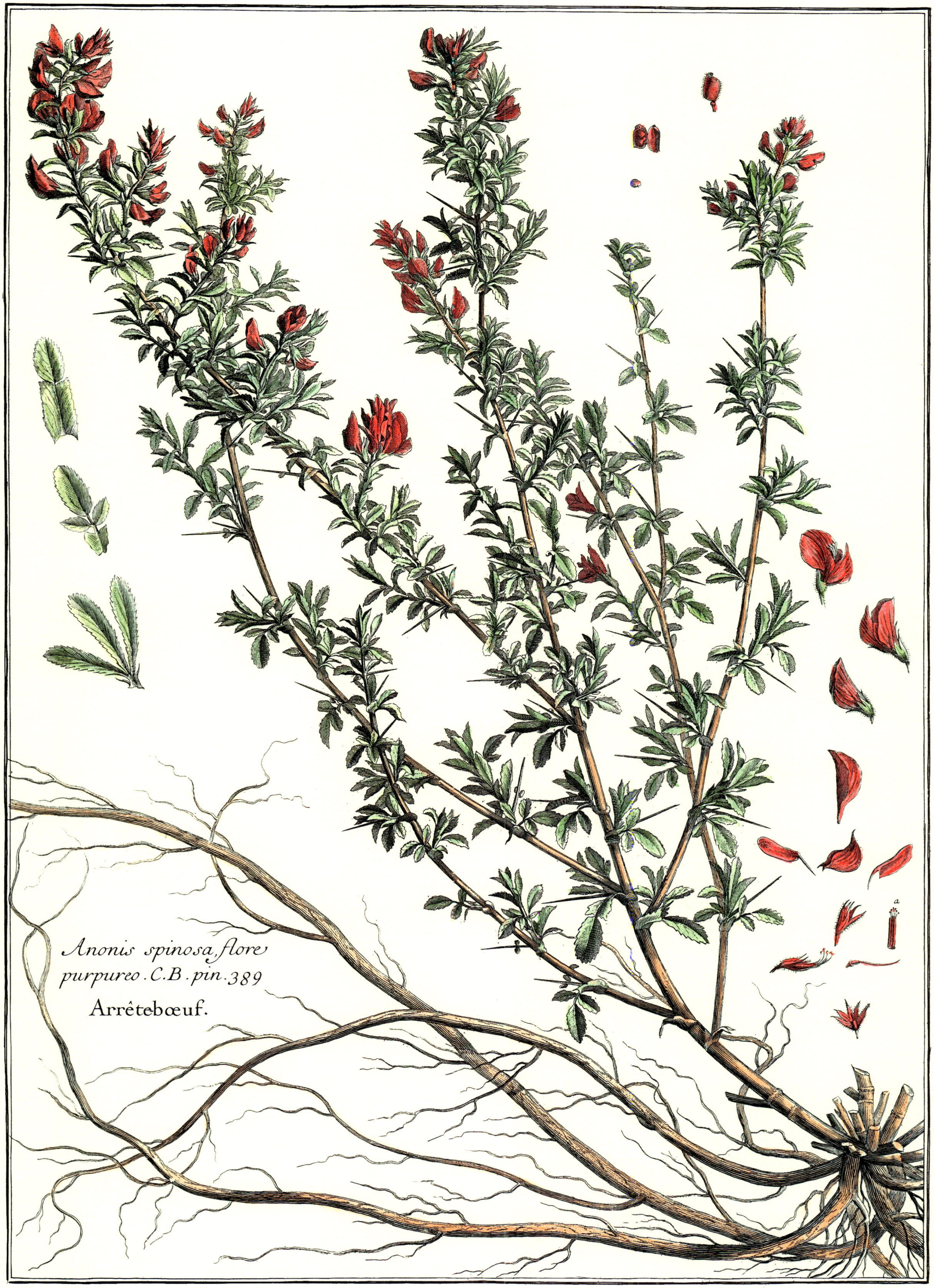

Anonis spinosa, flore purpureo. C.B. pin. 389

Arrêtebœuf.

Anagallis monelli L.

PRIMULACEAE

The species shown here is only adventitious in France, where it grows in the Aude region. It is found in dry, open habitats in Spain, Portugal, Morocco, Sardinia and Sicily. It is a hardy perennial, similar to the scarlet pimpernel, but red or blue in the case of *Anagallis arvensis* L. and blue in *Anagallis foemina* Miller. The French common name *mouron* probably comes originally from Old Dutch but is misleading as it is also applied to other species, in particular to *mouron blanc* (the common chickweed), *Stellaria media* L. Villars, which grows in the same situations. The scarlet pimpernel, a member of the primula family (like the primrose, the soldanella and the cyclamen) is completely glabrous except for the flower. The corolla is formed by five entire petals joined at the base and its fruit opens in the same way as an old-fashioned trinket box. The *mouron blanc*, however, is a member of the family Caryophyllaceae (like the carnation, the catchfly and the soapwort), and has a line of hairs along one side of the stem, switching sides between the nodes. The flowers have five white petals so deeply bifid that there appear to be ten, and the fruit opens into six deep valves. When it is not in flower, the line of hair on the chickweed is a reliable feature by which it can be identified.

As if to add to the confusion, both these *mourons* were once known in French as *morgelines*. This was because the first of them was believed to kill hens (*mort* meaning death and *geline* or *galine* being old French words for hen, which generated the Latin genus name). In the case of the *Stellaria media*, the *mor* in *morgelines* refers to the fact that hens are very keen on this plant (*mordent* = bite), commonly known in French as *mouron des oiseaux*.

For many years common chickweed was eaten raw as a salad leaf or cooked in *potherbes*, in omelettes, or chopped finely and sprinkled on to boiled potatoes tossed in salt and butter. Some connoisseurs, including some master chefs, are very keen on it and promote it with enthusiasm.

It is very important not to confuse this plant with *Anagallis*, which is toxic because of the hemolytic saponins it contains. However, *Anagallis* has been much used in medicine. Dr Leclerc, the celebrated twentieth-century herbalist, used it successfully to treat, among other things, pulmonary sclerosis following tuberculosis. The problem had been aggravated by congestion due to flu and had deteriorated into mucopurulent bronchitis. A cure was achieved by using liquid extract of pimpernel and Garus's elixir. Leclerc also achieved remarkable results by applying plant-based poultices to ulcers and wounds.

Yet none of this had anything in common with the way it was used by Dioscorides in the first century AD, as recounted by Daléchamps in 1615: 'Dioscorides states that both types of *Mourron* [author's note: referring also to the yellow pimpernel, *Lysimachia nemorum*] have this virtue, that they ease and prevent the development of inflammation. They draw out splinters and contain corrosive ulcers. The juice of this plant removes phlegm from the brain when used as a gargle and if inhaled through the nose. It eases toothache if placed in the ear on the side where the tooth is not hurting. Used with Attic honey, it cures eye disorders and sight problems. Taken as a beverage with wine it is beneficial for those bitten by snakes; also for kidney problems and for those with liver disorders (in the past it was also used to treat those with dropsy). Finally, Dioscorides writes, in agreement with the opinion of others, that the *Mourron* which has a blue flower brings back what has been deeply swallowed and that the red-flowered plant aids digestive descent when used as a rub. From which it would seem that these two plants are different not only with regard to their flower but also with regard to their effects.'

Anagallis monelli L. Engraving : Nicolas Robert
National Museum of Natural History, Paris.

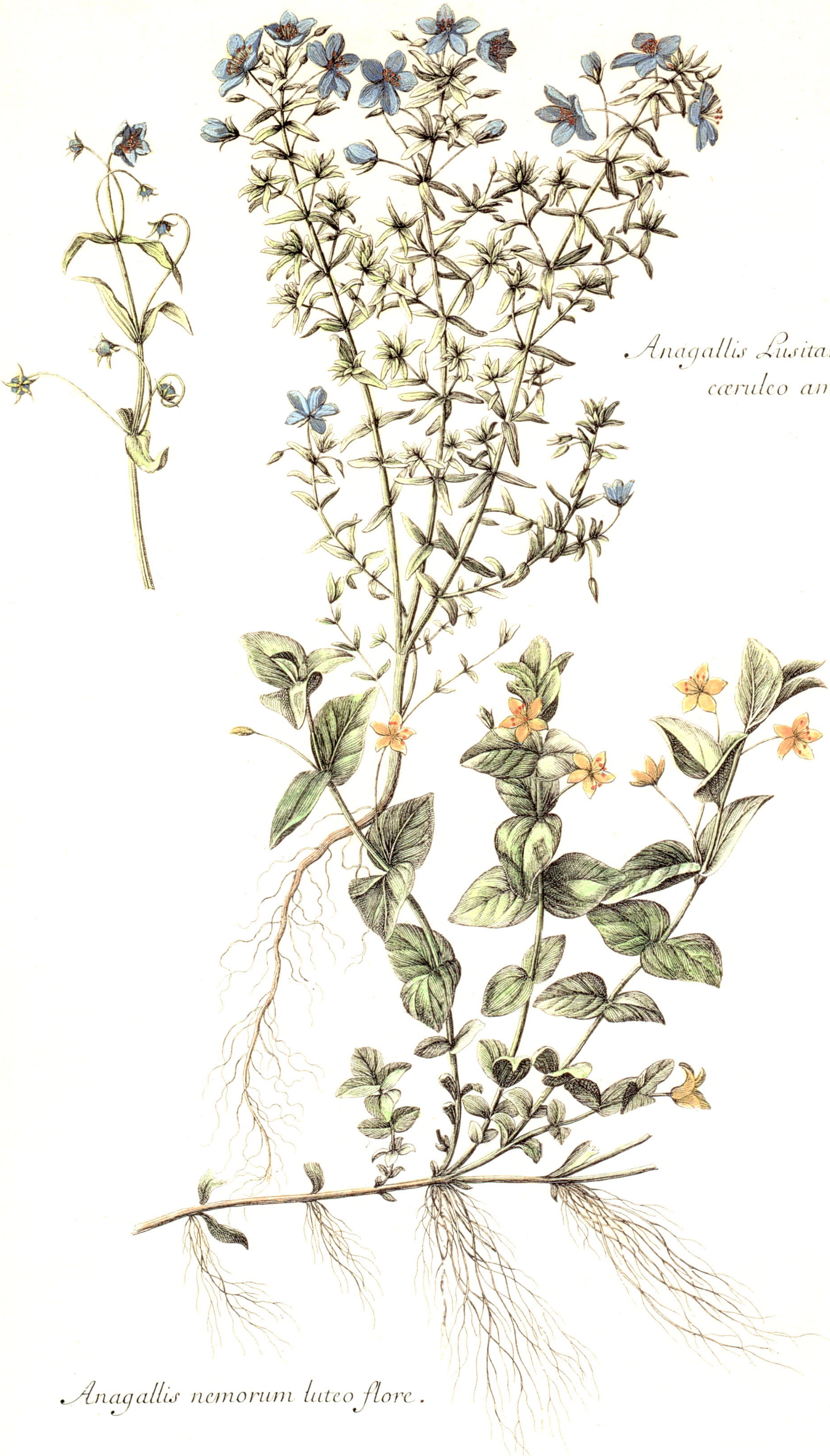

Anagallis Lusitanica flore cæruleo amplo.

Anagallis nemorum luteo flore.

N. Robert in. f.

Lysimachia nemorum L.

PRIMULACEAE

The Latin name *Anagallis nemorum luteo flore* can be literally translated as *mouron jaune* (yellow pimpernel), and is known in French as *lysimaque des bois* (the wood pimpernel). Whether upright in habit or creeping, *Lysimachia* is a graceful, charming plant. All the species are attracted by damp situations, brightening these up with their yellow flowers (white in the case of *L. ephemerum* L.). The yellow pimpernel prefers woodland floors while the other species are found in meadows and alongside streams and rivers. Mathias Lobel (1538–1616) mentions a *mourron jaune* – a testimony to his keen sense of observation as *Lysimachia* and *Anagallis* both belong to the primula family. In 1570, his friend Pierre Pena writes: 'In regions beyond the sea, there is a *Mourron jaune* that is handsome and rare and which is quite commonly found among woods and in shady spots in England. It flowers in June and July, and produces yellow flowers. It likes shady spots but for the rest resembles the common *Mourron*.'

According to most botanists the plant was named in honour of Lysimachus, king or son of the king of Sicily, or possibly after a doctor named Lysimachus. Daléchamps (1615) helps to clarify the origin of the word: 'This plant is named in Greek ... and also in Latin after King Lysimachus who invented it, as Pliny tells us, or for bringing an end to disputes and battles ... Indeed, this plant is of such virtue that it can quell the ferocity of animals and their conflicts and discords, as Pliny bears witness, describing it briefly in this manner: King Lysimachus was the first to invent *Lysimachia* and therefore it bears his name ... This herb has such properties that if placed below the yoke of two draught animals who do not wish to work together it will render them peaceful. Dioscorides describes it a little more fully. *Lysimachia*, known by some as *Lîtron*, produces stems the height of a cubit and more, that are branching and thin, from the nodes of which emerge leaves similar to willow leaves and astringent to the taste. Its flower is yellow or the colour of gold. It grows in marshy places and around water.'

Anatole Bailly includes the adjective *lysimachos* in his *Dictionnaire grec-français* (1894). It comes from *lysos* (to destroy), and *machos* (fight). A King Lysimachus did indeed exist. He fought alongside Alexander the Great who bequeathed him the country of Thrace where he built the city of Lysimachia. However, he spent so much time waging war that it is hard to imagine how he found the time to make use of the peace-giving *lysimachia*!

Not all sixteenth-century botanists were in agreement on the determination of genera. Anything that was tall and grew alongside water was called *lysimachia*. From ancient times, this plant was confused with purple loosestrife, at least by Pliny the Elder (23–79 AD) who spoke of a 'purple-coloured flower', though this was challenged by Dioscorides (first century AD). The German Rembert Dodoens (1517–85) considered loosestrife to be *Vraye Lysimachie* (true *lysimachia*) and called the hirsute epilobium *Lysimachie purpuree seconde*. This plant was named *Autre Lysimachie gousse* by Lobel as well as *Filus ante patrem*, as herbalists thought that the fruit appeared before the flower. It was in fact the long, slender capsule which contains the ovary and at the end of which the flower of epiboliums grows.

The Italian botanist Matthiole (1500–77) got around the problem by calling the yellow pimpernel *Lysimachie de Matthiol*, and purple loosestrife *autre Lysimachie*. *Veronica anagallis-aquatica* (the blue water speedwell) was named *Lysimachie bleue* and the common figwort, *Lysimachia galericulata, de Lobel.*

All of which was done in a polite and harmonious manner, not requiring the application of *lysimachia*!

Lysimachia nemorum L. (detail; see p. 69). Engraving : Nicolas Robert
National Museum of Natural History, Paris.

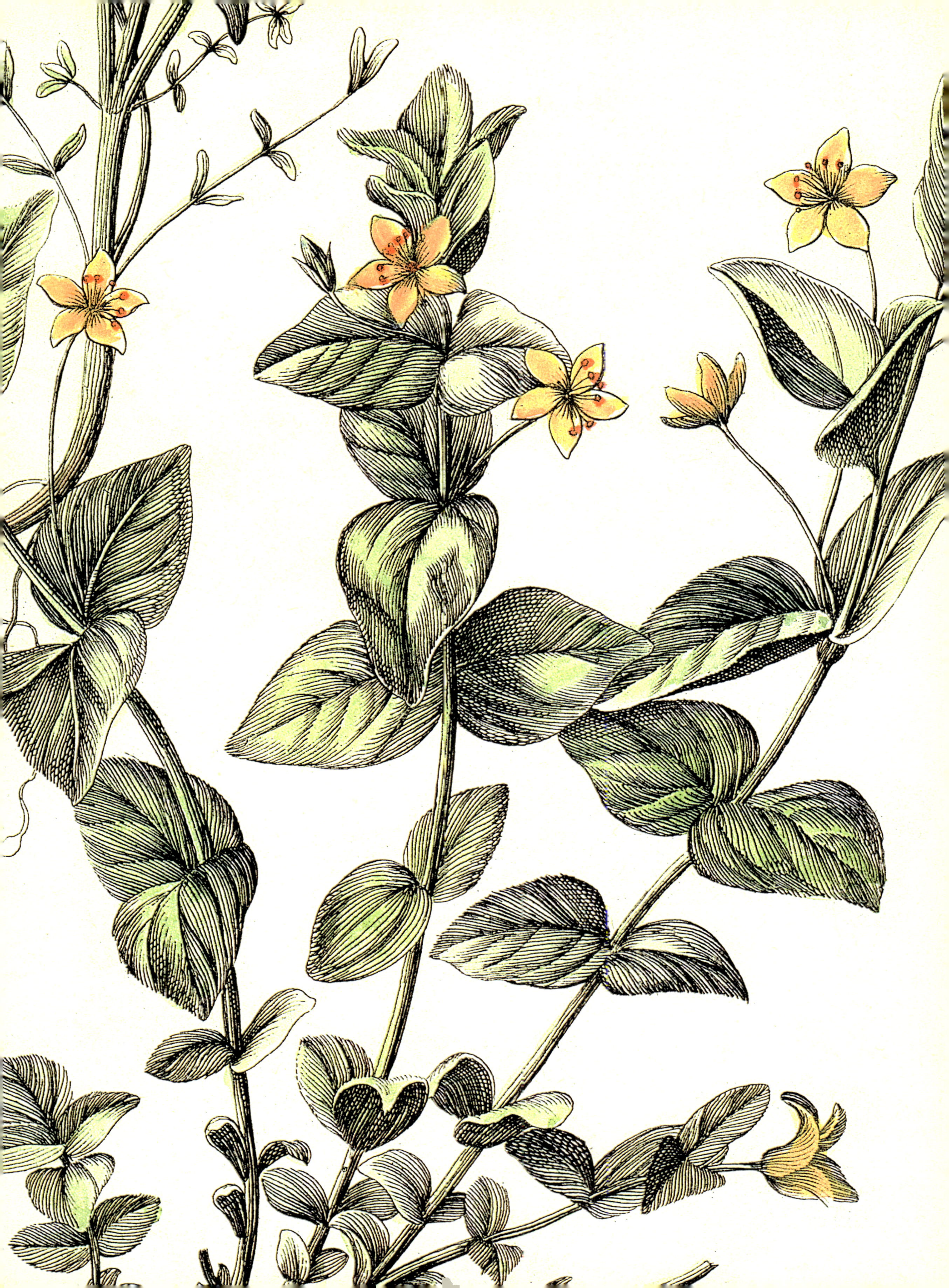

Lunaria annua (L.)

BRASSICACEAE (EX-CRUCIFERAE)

Everyone knows or has heard of honesty. It decorates florists' and other shop windows and is one of the most sought after plants for dried flower arrangements. It can be seen in gardens at different stages of development, but those who do not grow it will probably only be familiar with its seed heads, often unaware of the name of this plant with its beautiful flowers. The translucent partition or septum that divides the two valves of the fruit (known as a silique in the crucifer family) is reminiscent of a full moon with its rounded form and luminosity, hence the name *lunaria*. This partition that divides the 'pod' into two parts is what distinguishes crucifers from other families such as the papilonaceae and legumes. And the French common name *monnaie-du-pape* (pope's money) reflects the general impression people have of the pope's extraordinary wealth.

Two species of the genus *Lunaria* grow in France: *annua*, the plant illustrated here and which was once known as *biennis* (which seems more appropriate); and *redivivia* L., a hardy species.

The first of these is commonly found on roadsides, in well-lit habitats and on wasteland in warm, temperate climates. As with many crucifers, its flower head of lilac-purple flowers, keeps growing taller, leaving behind fruits in the form of circular siliques at all stages of maturity. The second, which has more fragrant flowers and more elongated siliques, grows in damp situations – meadows at the foot of hills, brightly lit woodland clearings, etc. – in cooler regions. There is also another species not found in France, *Lunaria telekiana* Jav., which is native to north-eastern Albania.

The flowers are visited by many insects, including bees and butterflies. And the plant plays host to at least one caterpillar, that of the orange-tip butterfly, *Anthocharis cardamines* L.

Daléchamps (1615) has this description of the plant, which he calls the 'large-leaved stock': 'Since the flowers of this plant closely resemble those of Violets and Stocks, Herbalists have counted it among the Stocks, calling it *Viola latifolia* or *peregrina*. Most modern ones call it *Lunaria*, and some *Lunaria Graeca* ... in Brabant it is called *Percininkbloemen*, that is to say 'money flower', because its pods closely resemble coins ... after the flowers come large, smooth, almost round pods, composed of several membranes, thin like small blades and glossy, each containing four or five seeds, and are plain, broad, brown in colour, almost in the shape of a split moon, between which there is a membrane, making this the third ... Pena says this plant grows in the mountains of Provence and on the border with Piedmont; but in France, Flanders and England ...Pena and Dodon say that there is another plant similar to this one, which has large leaves and flat, circular pods, a little smaller than the former, tougher and shorter stems, paler flowers, roots that do not die in winter, and therefore it lasts longer ... Lobel says that they are detergent and have mild heat-giving properties and bring on urination, as do rampions. They are eaten in salads like rampions and other plants.'

Lunaria annua L. Engraving : Abraham Bosse
National Museum of Natural History, Paris.

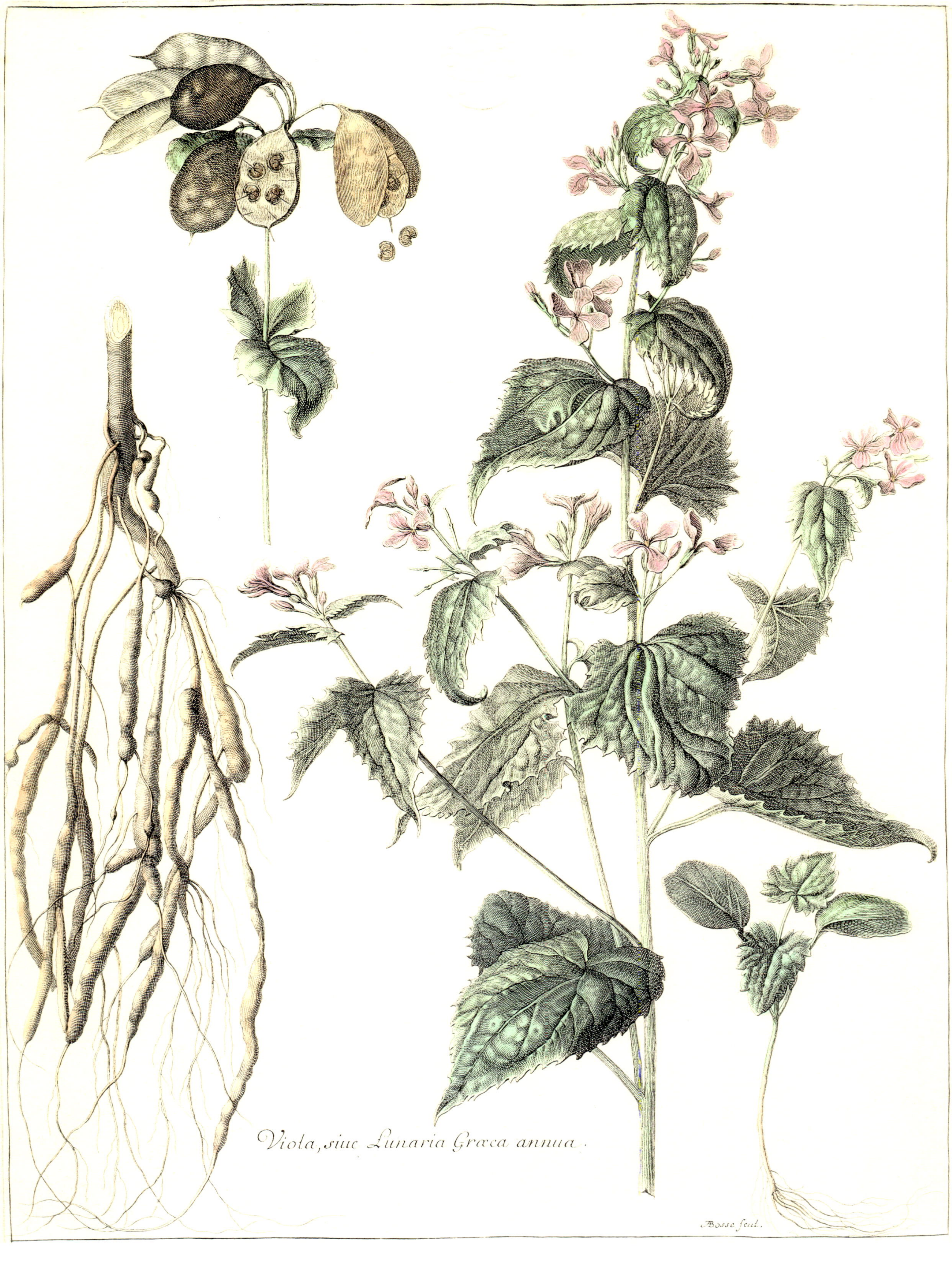

Viola, siue Lunaria Græca annua.

Hesperis matronalis L.

BRASSICACEAE (EX-CRUCIFERAE)

It seems that during the Middle Ages great confusion existed in the crucifer family between wallflowers and various types of stock and rocket. Indeed it is hard to know if what Dioscorides called *Leuconium* or white violet in the first century AD was what Pliny referred to as *Hesperis*.

Daléchamps (1615, book XII) mentions the *violier des jardins*: 'The great *violier de mer*, which is quite common along the Languedoc coast, has long, narrow leaves, lying flat to the ground, serrated like those of chicory or rocket, large, whitish and of a size with those of *violiers des jardins* ... Some take them to be Theophrastus's *Hesperis* ...' But was he talking about gillyflowers or about *Hesperis*, our sweet rocket? Could the word *violier* have been transmuted into *violette* (violet) to give '*violette de Damas*' (dame's rocket)? Whether this is the case or not, we do know that *violette de Damas* became the *violette des dames* that Rembert Dodoens (1517–85) translates as *matronalis*.

Julienne (sweet rocket) grows happily in shady areas in central France and mountainous regions, but is less common in open country. It is found almost everywhere throughout the northern hemisphere. For example, it has colonised the shores of St Laurence Bay – along the river from Blanc-Sablon, Natashquan, Sept-Iles, Baie-Comeau, Quebec, Montreal, Rivière-du-Loup and Gaspé to the Isles of Mingan and Anticosti.

Cazin (1868) says that it was successfully used to treat cases of scurvy and pulmonary catarrh; however, it has been little used in medicine. Although there are no specific studies on sweet rocket, it is believed that it may be the presence of sulphur heterosides, found in certain crucifers and irritant to the gastric mucosa, that provokes an increase in bronchial secretion by vagal reflex. This is true of another well-known crucifer, the hedge mustard which has a healing effect on the lungs and vocal cords.

The origin of the French common name *julienne*, which appeared in 1677, is not known. The plant has a number of other vernacular French names including *aragone*, *cassolette*, *beurrée*, *girarde*, *giroflée musquée* and *hespéride des jardins*. In English it is known mainly as dame's rocket, sweet rocket, and damask.

The name *Hesperis*, from the Greek *hesperos* (evening), probably relates to the fact that the fragrance it gives off is strongest at the end of the day. In fact, everything related to evening was located in the west, where the sun sets, and at least two countries were once known as *Hesperia*: Italy by the Greeks and Spain by the Romans. As for the mythical garden of the Hesperides, the Ancients placed this on the frontier of human knowledge at that period, that is to say just off the west coast of Africa.

The Hesperides, the 'nymphs of the evening' were, with the help of a dragon, responsible for guarding the garden of the gods where the trees bearing the golden apples sought by Hercules grew. In one of the many variations of this myth, the Greek hero kills the dragon and steals the golden apples. In desperation the three nymphs, Aegle (the dazzling), Erytheia (the red one) and Arethusa (the setting sun) transform themselves into trees – an elm, a poplar and a willow – under the shade of which the Argonauts later sheltered. The dragon became the constellation of the Snake.

Hesperis matronalis L. Engraving : Nicolas Robert
National Museum of Natural History, Paris.

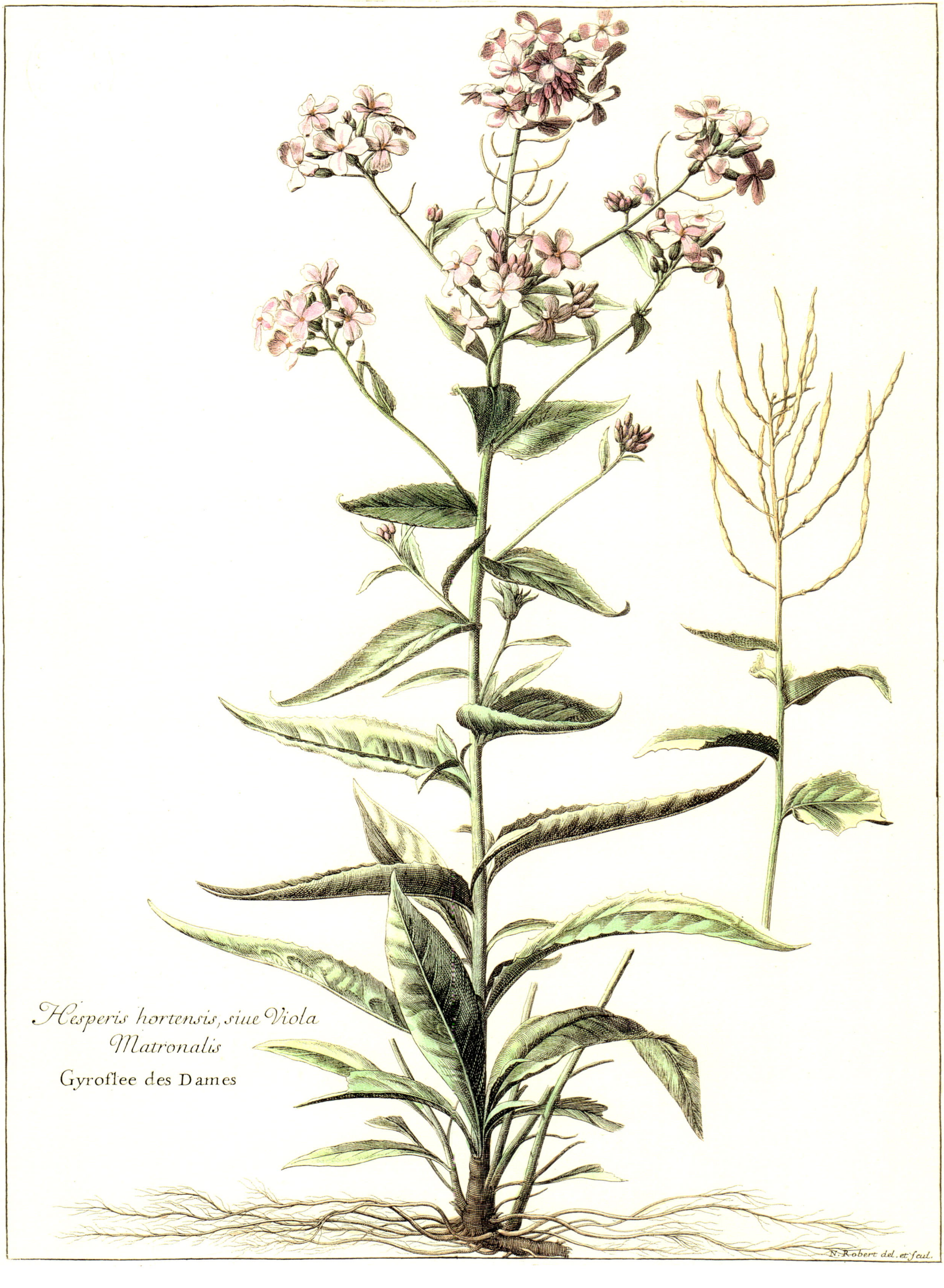
Hesperis hortensis, siue Viola Matronalis
Gyroflee des Dames
N. Robert del. et sculp.

Alliaria petiolata (M. Bieb.) Cavara & Grande

BRASSICACEAE (EX-CRUCIFERAE)

This annual or biennial plant is commonly found throughout France, growing in gardens as well as on embankments, wasteland, below hedges and at the foot of walls – in fact in most areas where the soil has been disturbed. The basal leaves are broad, reniform and crenate, with a long slender petiole, while the upper leaves are oval and heart-shaped, ending in a point, in a similar way to those of nettles and balm. Its terminal flowers seem unending: the small clusters of pretty, white flowers with four petals in the form of a cross, always found at the top of the flower stalk, leave behind them fruit in the form of long, upright siliques. So, by the end of the flowering period, all stages of development can be seen, from the flower bud through to the fully mature seed pod, with its characteristic oblong, striated seeds.

The *alliaire officinale* (garlic mustard) contains substances with the odour of garlic and mustard found in many crucifers such as hedge mustards, mustards, cardamines, winter cresses, turnips, and cabbage. The English name 'garlic mustard' is very apt. The term 'must' was first used to describe a condiment made from unfermented grape juice and was subsequently applied to the plant used in this condiment (1278) and then to its seed. Some etymologies link it to *moût ardent* (burning must) because of its spicy taste.

The leaves of the garlic mustard were consumed for many centuries and continue to be eaten in the mountainous regions of the south of France. A handful of leaves and flowers, both edible – and delicious – together with some rampions, French scorzonera, perennial lettuce and wild corn salad make a salad fit for a king!

The term *officinale* refers to its medicinal properties: it has stimulant, bechic, diuretic, vermifuge and other properties. In the Cevennes region and in the southern Alps, it was once common practice to take a 'depurative' in spring. This was done over several days using a number of different plants of varying bitterness, such as the wall germander, burdock, the common centaury, the white horehound and garlic mustard. Garlic mustard broth was considered extremely beneficial for skin conditions.

At the time of Daléchamps (1615), the *Herbe aux aulx* – as it was then known – was believed to have many virtues: 'The whole plant has a rather warm, damp taste. Because of which women, principally English women, often mix the crushed leaves into sauces, being of the firm opinion that they are good for preserving good health throughout the year. Its properties are similar to those of hedge mustard and rocket, for which reason Pliny believes it to be good for coughs. The warm and dessicative qualities of this plant suggest that it can ease gross humours and act on viscous ones. It is said that if the seed is ground into a plaster form and applied to the womb it can cure women who have suffered haemorrhage and restore them to health.'

The term 'crucifer' means 'one who carries the cross', from the Latin *crux* (cross) from which many other words are derived, such as crucial, crossing, crusade and cruise (*croisière* in French); the word *croisière* was first used to mean 'things that pass by each other' (1344) and was adopted for maritime use to describe the waters where vessels pass one another (1678). Of course its current meaning has nothing in common with the perilous expeditions described in the foreword of this work (pp. 14 and 18).

Alliaria petiolata (M. Bieb.) Cavara & Grande. Engraving : Abraham Bosse
National Museum of Natural History, Paris.

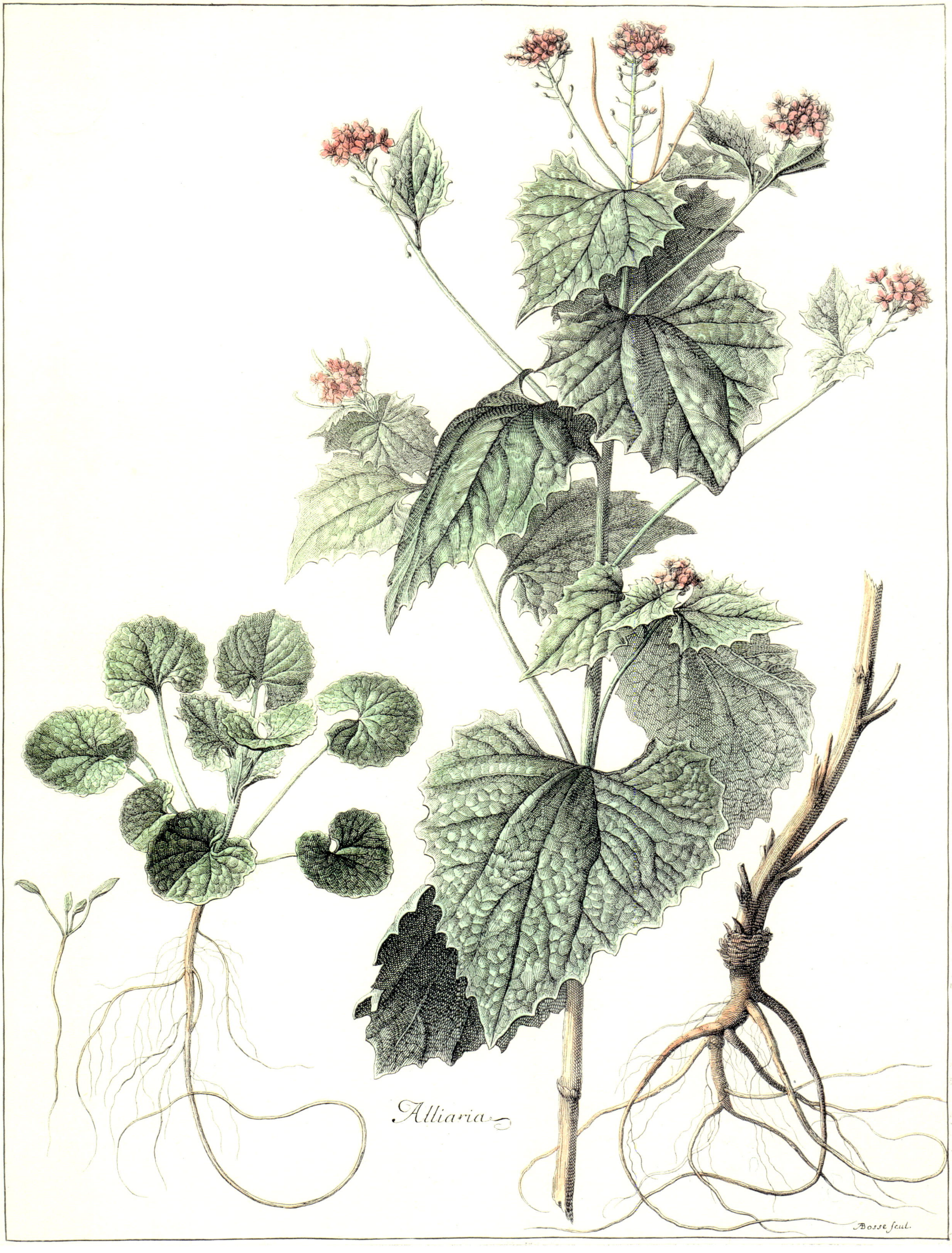
Alliaria
A.Bosse sculp.

Aesculus hippocastanum L.

HIPPOCASTANACEAE

If the colour of this engraving is faithful to its model, this is the *marronier rouge*, *Aesculus x carnea* Hayne, known in English as the red chestnut or red horse chestnut, a hybrid produced by the crossing of *Aesculus hippocastanum* L. and *Aesculus pavia* L., discovered in the early twentieth century. Where the flower is predominantly white, the species is *hippocastanum*. Both species are used in the manufacture of Bach Flower Remedies.

It is curious that the term *Inde* (as in Indian horse chestnut) has continued to be applied to this chestnut, since it originated in the mountains of Turkey and Greece and that it was never mentioned by the ancient Greeks. Also curious is the fact that in 1737 the great Swedish botanist Linnaeus retained the Latin genus name *Aesculus*, which referred to a green oak with edible acorns, whereas horse chestnuts are not edible and the two trees have almost nothing in common.

The term *Hippocastanum* (horse chestnut) may come from its use by the Turks as a stimulant for weary horses or from the horseshoe-shaped scars left by the leaves on the current year's branches.

The first to describe it was the Italian botanist Matthiole (1500–77) who, in 1565, was given a branch of this chestnut bearing leaves and fruits by the Flemish physician Quackelbeen. The latter was in the service of the ambassador of the Holy Roman Empire in Turkey, Augier Ghislain Busbeck, whose name is associated with the tulip. Eleven years later, Charles de l'Écluse, known as Clusius (1526–1609), who at the time was in charge of the imperial gardens in Vienna, received some seeds from the ambassador to Constantinople and planted them. He gave a description of it in 1581.

In Paris, the first horse chestnut was not planted until 1615. Brought back from Constantinople by the French botanist Bachelier, it adorned the gardens of the Hôtel de Soubise in the Marais area of the city. This tree in fact survived until 1840. A second horse chestnut was planted in 1650 in the Jardin du Roi (now the Jardin des Plantes) but only survived for seventeen years.

The term chestnut applies both to the large, unpartitioned seeds of the sweet chestnut and to those of the horse chestnut, the latter being inedible due to the large amount of saponines it contains, which make the kernel highly toxic, even fatally so in high doses. When used externally, these same saponines have properties similar to those of saponin, acting as fat emulsifiers. The pulp of the horse chestnut has therefore been made into a paste and used as soap for washing textiles, as a hand wash, and in glue, alcohol and candles.

Medicinally, intract of horse chestnut is a very ancient remedy and was used to treat varices and haemorrhoids and as a venous tonic as it contains catechols. Saponosides have an anti-inflammatory and anti-oedemic effect, particularly in cases of phlebitis. The bark contains a coumarin called esculoside which increases capillary resistance.

Now very common, horse chestnuts are often planted as ornamentals in parks and avenues because of their beauty, their rapid growth and the generous shade they provide. They are also popular with bees as they contain an abundance of quality nectar. Children too are fond of horse chestnuts, using their leaves and fruits to fashion small figures and animals and by the English in the well-known game of conkers.

The French word *marron* probably originates in the pre-Roman root *marr* (pebble, stone), which gave rise to the word *marelle* (hopscotch), the playground game that involves throwing a small stone.

Aesculus hippocastanum L. Engraving : Abraham Bosse
National Museum of Natural History, Paris.

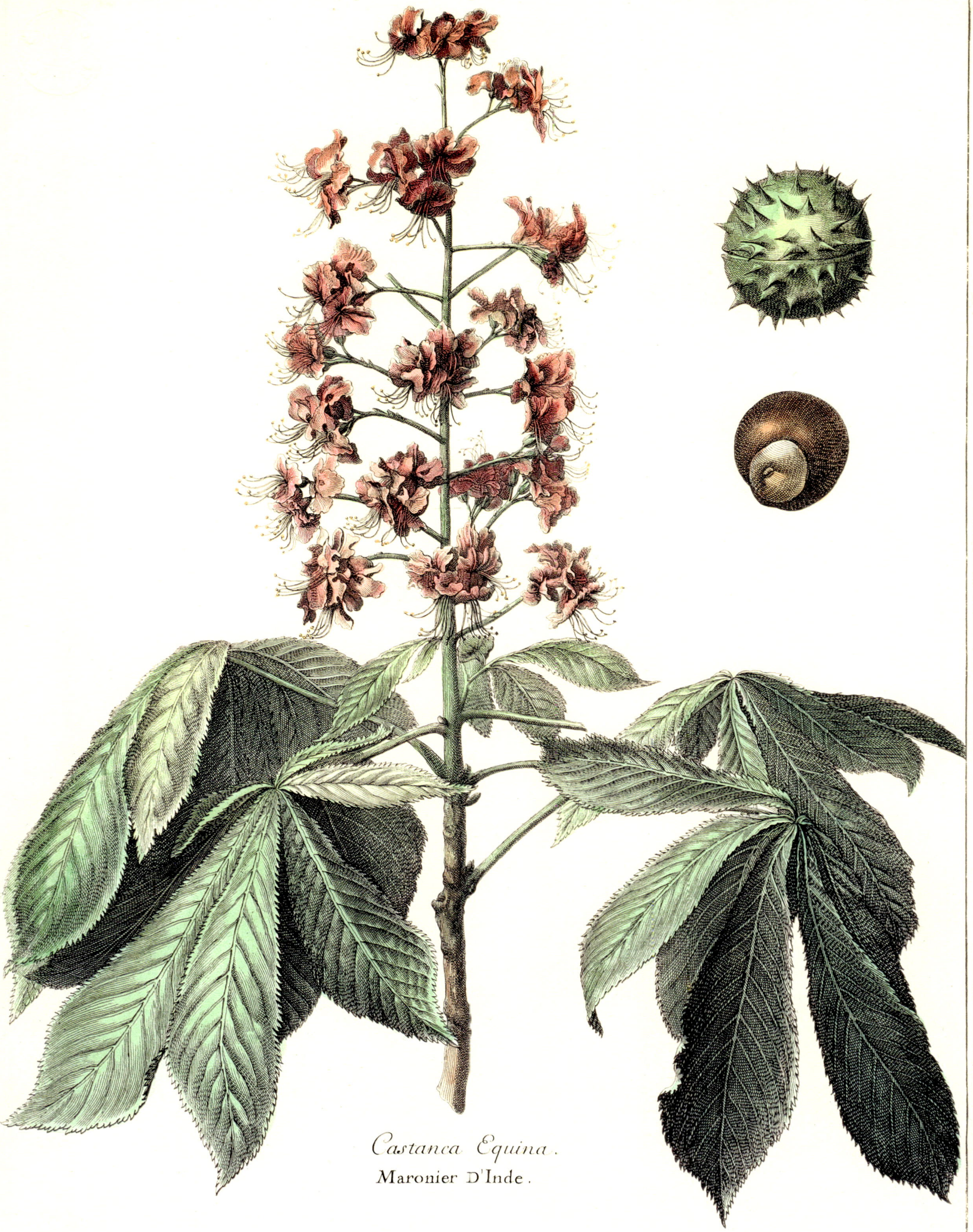

Castanea Equina.

Maronier D'Inde.

A. Bosse sculp.

Dictamnus albus L.

RUTACEAE

Some plants have properties that appear to smack of pure fantasy. This is the case with the fraxinella (also known as the burning bush or gas plant). A large plant with superb flowers, in summer *Dictamnus albus* forms clusters of beautiful pink and whitish flowers. The leaves on the stems consist of leaflets, similar to those of the ash, hence the French name *fraxinelle* and the vernacular name *petit frêne* (little ash).

In Antiquity there is a reference to a *dictamnus*, probably the dittany of Crete, *Dictamnus cretica*, an oregano (*Origanum dictamnus* L.). This small labiate plant with silvery leaves grows on cliffs and is associated locally with love because in the past young men would risk climbing the cliffs to pick this fragrant plant and present it to their loved one! 'Oregano', which means mountain beauty, comes from the Greek words *oros* (mountain) and *ganos* (joy, happy appearance).

Charlemagne mentions it in his *Capitulare de villis vel curtis imperialibus*: *'Volumnus quod in horto omnes herbas habeant, id est: lilium, rosas, fenugrecum, costum, salviam, rutum ... diptamnum, sinape, satureiam, sisimbrium ...'* ('We should like them to have plants of all kinds in their gardens, for example, lilies, roses, fenugreek, costmary, sage, rue ..., dittany, mustard, winter savoury, hedge mustard ...'). John Gerard (1545–1612) confirms that this refers to dittany of Crete, though it is surprising that it was possible to grow this plant on French soil because of the cold.

The presence of fraxinellone and dictamine endow this plant with a number of properties, and the fresh root, or its alcoholic tincture, were used in the past as a stimulant, anti-spasmodic, vermifuge, diuretic, febrifuge, expectorant and emmenagogue. The plant has now virtually dropped out of medical use.

Its leaves contain cineol, isoquercitrine, rutin and anethole and in Siberia were dried and used to prepare a tea-like drink said to aid digestion. However, various writers warn against this use, believing it to be dangerous.

The leaves of the fraxinella are punctuate, rather like those of St John's wort, with minute transparent pockets containing methylchavicol and anethole. The stem too is covered in brownish glands. On warm summer evenings, its lemon-fragranced essential oil permeates the air and can easily be ignited with a match.

The celebrated physician François Joseph Cazin has this to say about it in 1850: 'In warm weather, particularly around dusk, this aromatic vapour condenses and will ignite if a candle is held to it and presents the curious spectacle of a luminous halo which leaves the plant unharmed.' Some writers have even claimed that this essential oil can ignite spontaneously – which earned the fraxinella the French name *herbe aux éclairs* (lightning plant) and the English name 'burning bush'!

Both its beauty and the unusual flammable nature of its vapour have made this plant much sought after as a garden ornamental. Specialist plant catalogues offer fraxinellas with white, pink and bright purple flowers. However, it is worth noting that for people with sensitive skin it can cause persistent burning and blisters if the skin comes into contact with the leaves, rather like the wild parsnip.

Dictamnus albus L. Engraving: Abraham Bosse
RMN, Louvre Chalcography Dept, Paris (modern print).

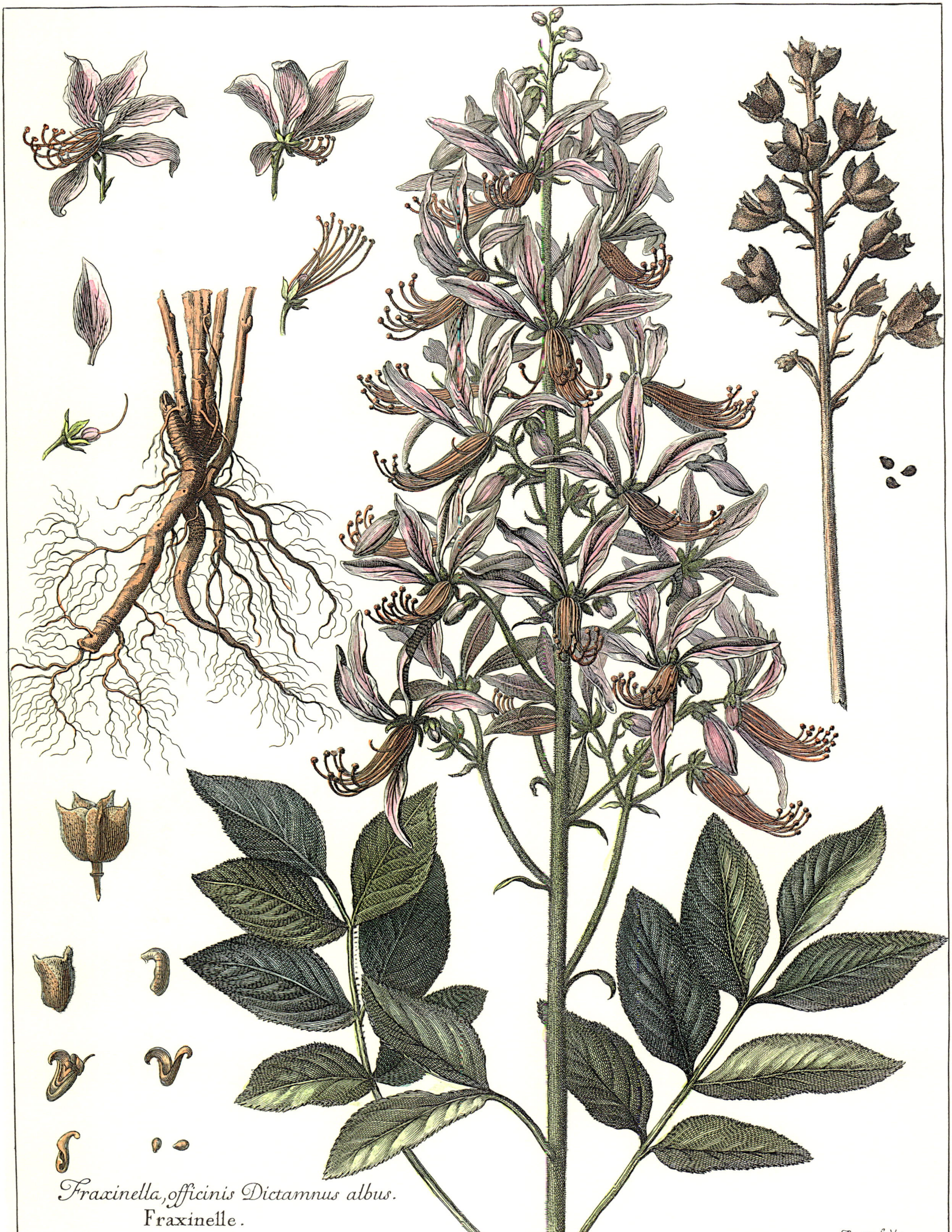

Fraxinella, officinis Dictamnus albus.
Fraxinelle.

A. Bosse sculp.

Coriandrum sativum L.

UMBELLIFERAE

Coriander is a plant that has been grown and picked for many centuries. It appears to have originated in the eastern Mediterranean but rapidly went on to be cultivated in and colonise the whole of the Mediterranean basin. It is mentioned in ancient Egyptian, Hebrew, Greek, Roman, Sanskrit and European texts. In France it appears as early as the eighth century in the *Capitulaire* of Louis the Pious and then in Charlemagne's *Inventaires* in 812. In cultivation since the Middle Ages, coriander is now found almost everywhere, though it prefers a warmer climate.

The name coriander comes from the Greek *koriannon*, which many etymologists associate with *coris* (the bed bug) because of its unusual smell. In the past, its fruit was classed as one of the 'four hot seeds' by apothecaries, along with fennel seed, green anise seed and caraway. These are digestives, carminatives (preventing the formation of gas while also facilitating the expulsion of gas already present), diuretics, anti-spasmodics due to the presence of linalol, and stimulants. Linalol is a non-phenolic, terpenic alcohol also found in many aromatic labiate plants such as thyme, savory, basil and creeping thyme.

In Ayurvedic medicine, coriander (*dhanyaka* in Sanskrit), with its *karma* effect, is considered to be an energizer, a diaphoretic (drawing out toxins through transpiration) and a carminative. Its *Rasa* flavour makes it mild, its *Virya* energy makes it refreshing and its *Vipaka* post-digestive action makes it pungent, thus rebalancing the three *Doshas*, that is to say the three fundamental elements of the human constitution: *Vata* (air and ether), *Kafha* (water and earth) and *Pitta* (fire and water). Those of strong *Pitta* constitution tolerate the fruit of the coriander well but should avoid other spices.

In his *Histoire générale des plantes* – written in Latin in 1585 and translated into French in 1615 – Daléchamps praises the virtues of coriander but goes on to reveal its drawbacks: 'Most often, apothecaries do not prepare the coriander from which they make dragées, but only use sugar and heat. It should not therefore be thought that this seed loses its pungency in the coldness of the vinegar, but rather by being dried. The coriander herb, and in particular its juice, is cold, so much so that in some places it is fatal, according to the testimony of Arabs, when taken as a beverage, as modern practitioners have seen from their experience with the "Hiurca" or "Yucca" of America, which is a plant that is good to eat and nutritious, yet its juice can kill a person.' A puzzling statement to say the least.

The French word *apothicaire* (apothecary) appeared in the form *apotecaire* in 1260, derived from the Greek *apotheke* (shop), which gave rise to the old Provençal word *botica*, which in turn produced *boutique*. Originally, apothecaries sold rare plants and spices – the origin of the French word *épicier* (grocer) – and then began to specialise as 'preparers and vendors of medications', entering the field of pharmacy (1620). Pharmacy comes from the Greek *pharmakon* (medicinal plant), or perhaps from *pherein* (which bears) – with its variant *pharein*, meaning 'plant borne by the earth' – and from *akeios* (plant remedy). And indeed, coriander (and other spices) can still be purchased in pharmacies as well as grocer's shops.

Coriandrum sativum L. Engraving : Nicolas Robert
National Museum of Natural History, Paris.

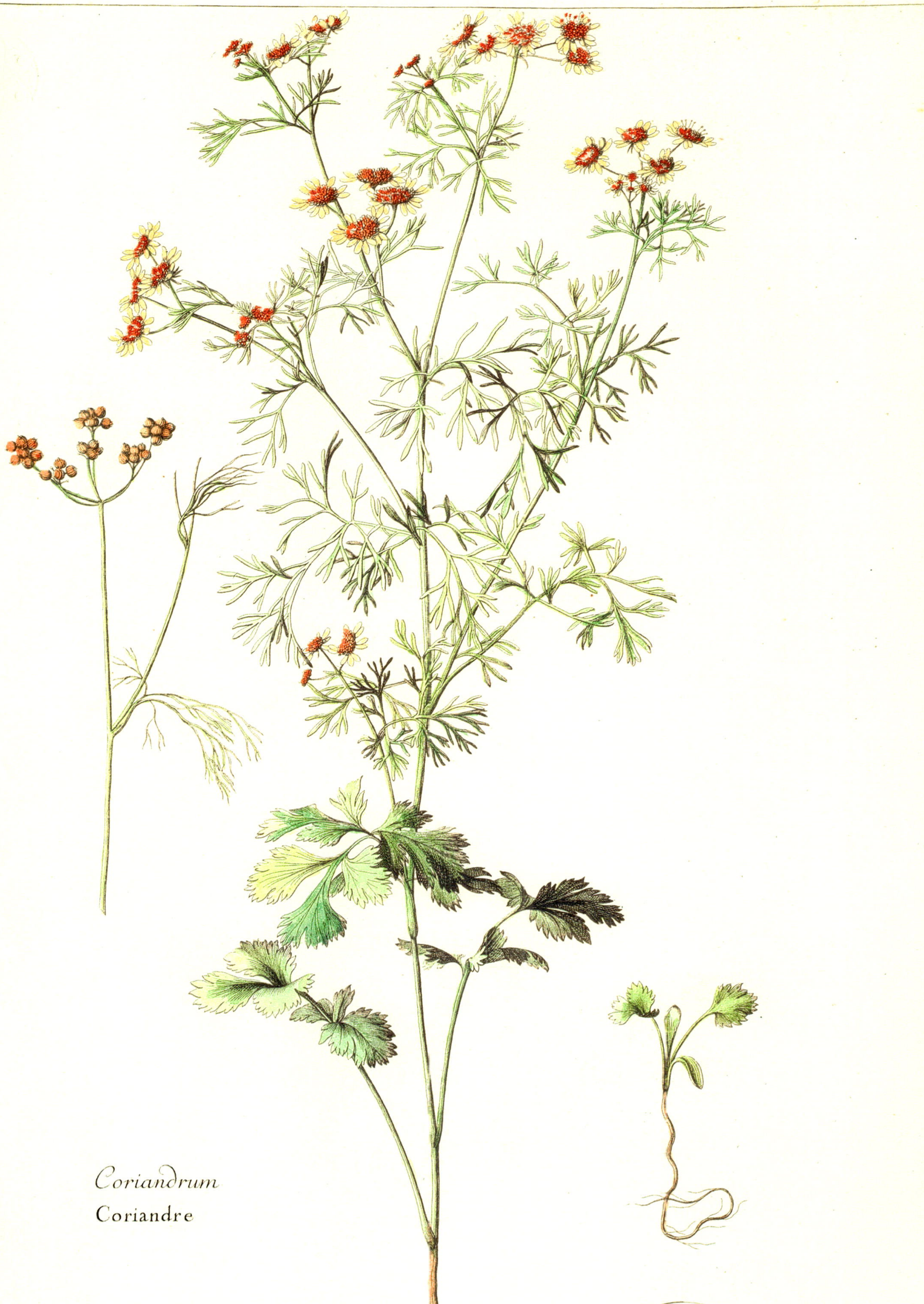

Coriandrum
Coriandre

N. Robert del. et sculp.

Gentiana acaulis L.

GENTIANACEAE

There are many species of the genus *Gentiana* living in mountain areas. The majority produce flowers of a splendid, luminous blue while the great yellow gentian is stunningly majestic.

Gentiana acaulis, the stemless gentian or Kock's gentian (ex-*Gentiana kochiana* Perr. & Song.) is of no medicinal use, nor is the pretty spring gentian, or indeed most of the other gentians. The yellow gentian, *Gentiana lutea* L., is the only one really used. Some species have been accepted as substitutes by Swiss and German pharmacopoeias, including *Gentiana asclepiadea* L., *Gentiana pannonica* Scopoli (originally from Hungary), *Gentiana punctata* L. and *Gentiana purpurea* L.

Gentius, king of Illyria (the northern part of the Balkans), is regarded as having discovered the properties of this plant. As a king he enjoyed only a very short reign, having started out badly in 172 BC by murdering his brother in order to gain the throne, and eventually came to a sticky end. His ally Perseus, king of Macedonia, abandoned him. Gentius was conquered by the Romans four years later and taken to Rome by the praetor Anicius. According to Pliny the Elder (23–79 AD), he was cured of a fever by the great yellow plant (or another species) and King Gentius's plant became known as the 'gentian'.

Around 50 AD, Dioscorides in his work *Materia Medica* – translated by Matthiole (1500–77) in 1572 – refers to the use of the yellow gentian to treat liver and stomach conditions as well as other disorders, a use confirmed by Galen (131–201 AD), Pliny, St Hildegard (1098–1179), Leonhart Fuchs (1501–66) and Hieronymus Bock (1498–1554), among others.

In France, where malaria was rife almost everywhere, it remained one of the few remedies effective against intermittent fever, also known as 'quartan fever' as attacks often tended to recur on the fourth day. In 1868, Cazin states: 'Before the discovery of cinchona, it was the most common remedy against intermittent fever: *Cortice peruviano nundum invento, sola gentiana febres quartanas expugnatas esse* (Boerhaave). It is sometimes combined with 'Peru bark' when the fever is particularly resistant ... Bitter rather than astringent, it has a beneficial effect on the action of cinchona.'

For many years, gentian root was used as a tonic for digestive and biliary function and as a febrifuge. St Albert the Great (1193–1280), a Dominican monk and alchemist, is believed to have developed the first recipe – for stomach treatment – of a medicine based on gentian extract. One of the plant's many chemical principles, amarogentioside, is the most bitter substance occurring in nature. The bitterness is still perceptible at a dilution of 1/58,000,000, whereas with gentiopicroside it becomes unnoticeable beyond a dilution of 1/12,000.

However, anyone thinking of picking gentian should be warned that there is a great danger of confusing it with the highly toxic white hellebore or false helleborine, *Veratrum album* L. (MELIANTHACEAE, ex-LILIACEAE). The leaves are very similar but differ in the way they are arranged: in the gentian they are opposed whereas in the helleborine they are alternate. Once the plants are in flower, confusion is no longer an issue: the tiered, yellow flowers of the gentian are well known, while the helleborine has a terminal spike of small, white flowers. The roots are also distinct, being short, thick clumps in the helleborine but thick and long (almost 1 metre/3 ft) in the gentian.

Gentiana acaulis L. Engraving : Abraham Bosse
National Museum of Natural History, Paris.

Gentianella Alpina verna magno flore.

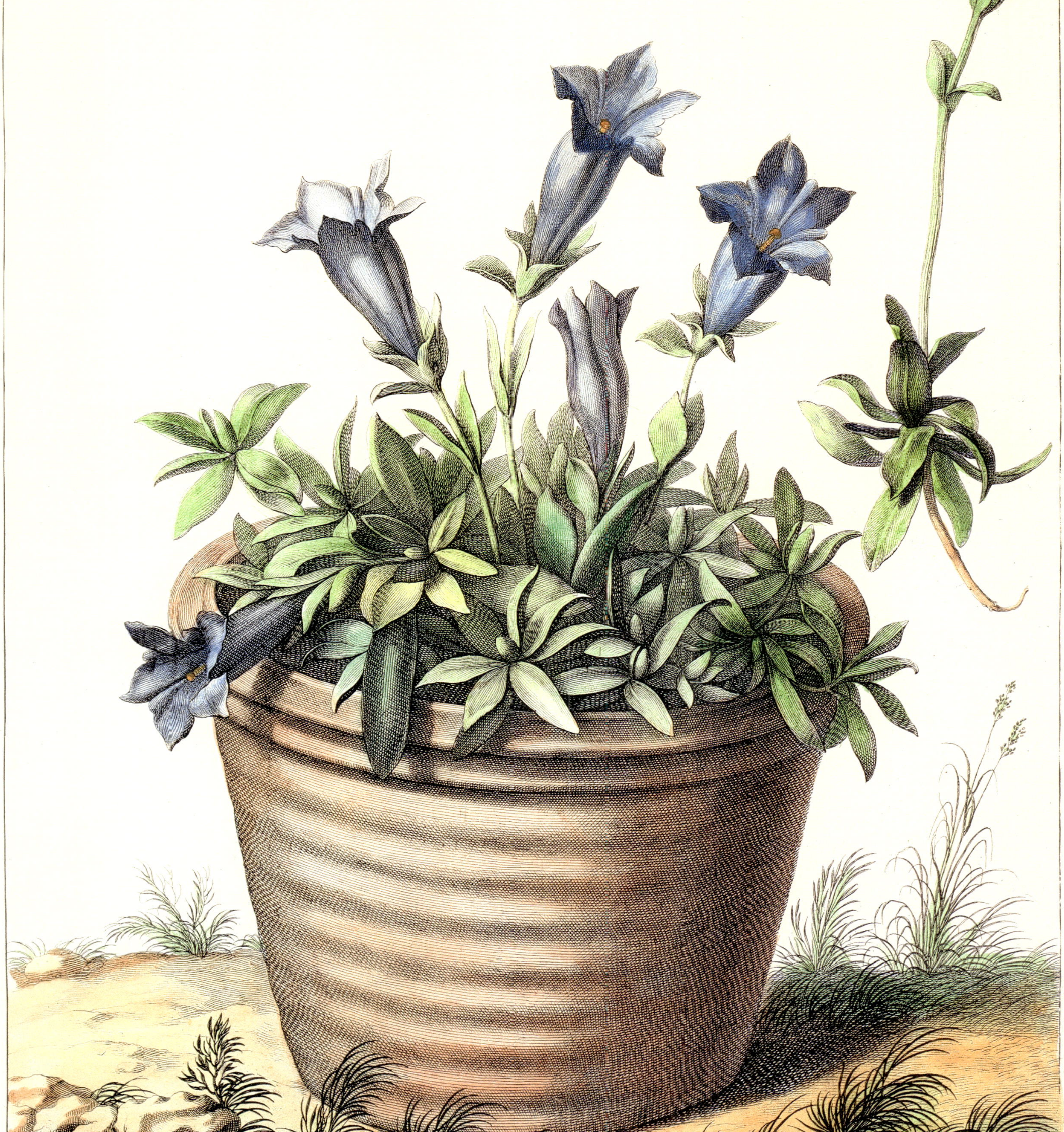

Mandragora officinarum L.

SOLANACEAE

Mandrake! The very name of this plant conjures up an entire period of the most unbelievable myths: up to the eighteenth century, many people believed that the mandrake grew under gallows, springing up from the sperm of hanged men, and that the scream it emitted when picked drove people mad!

Picking mandrake involved an elaborate ceremony – the person officiating placed wax plugs in his ears, drew three magic circles around the mandrake with a dagger, dug down into the soil to loosen the plant and then tied it to a dog. The animal was then frightened into running away, uprooting the plant as it fled. Its flight was soon interrupted as the mandrake's screaming would kill the dog on the spot! After all this, the plant was washed and dried and from then on it became like a living being, remaining loyal to its owner and bringing him fertility and prosperity.

The mandrake's forked roots, with their anthropomorphic forms, engendered beliefs that still persist. In the engraving shown here, Abraham Bosse appears to have added a personal touch, transforming the plant into a *homonculus* ('little man') – here in fact a woman – a being that some alchemists also endeavoured to create. A distinction was also made between 'male' roots and 'female' roots, though this has no botanical foundation as all these plants produce fruit.

In France the mandrake is not found growing in the wild but can easily be cultivated in gardens in the south of the country. It is found in the countries of southern Europe: Spain, Portugal, Italy, Greece; in North Africa: Morocco, Algeria and Tunisia; and in the Middle East: Israel, Jordan, Lebanon, Syria, Turkey.

There is some doubt about the origin of the name of this plant, though it may be derived from the name of a Greek physician, or from the Persian term *mendum gija* (the man plant). Towards the end of the twelfth century, it was known in French as *mandegloire*, then in 1436 as *maindegloire* and as *main de gloire* in 1752. Its English name, mandrake, will probably bring back memories to comic strip enthusiasts.

In 1484, Pope Innocent VIII issued a papal bull ordering the hunting down of witches. Persecution began in earnest following the publication by two Dominican monks of *Malleus Magnificarum*, a handbook describing witches and the methods to be used in order to identify them. Waves of arrests followed throughout Europe, mainly of women. In France, between 1560 and 1680, tens of thousands of ordinary working class women died under torture or were burned alive simply because of their knowledge of medicinal plants – and these punishments extended to their children if they were girls.

For centuries, mandrake was used in 'magic' concoctions, most often for purposes we would consider dubious today, but also for its hypnotic and analgesic properties. However, medicinal use of the plant was a dangerous practice as the root contains scopolamine and hyoscyamine, as do henbane, deadly nightshade and datura, all members of the family Solanaceae and which are the source of many other stories.

Mandragora officinarum L. Engraving : Abraham Bosse
RMN, Louvre Chalcography Dept, Paris (modern print).

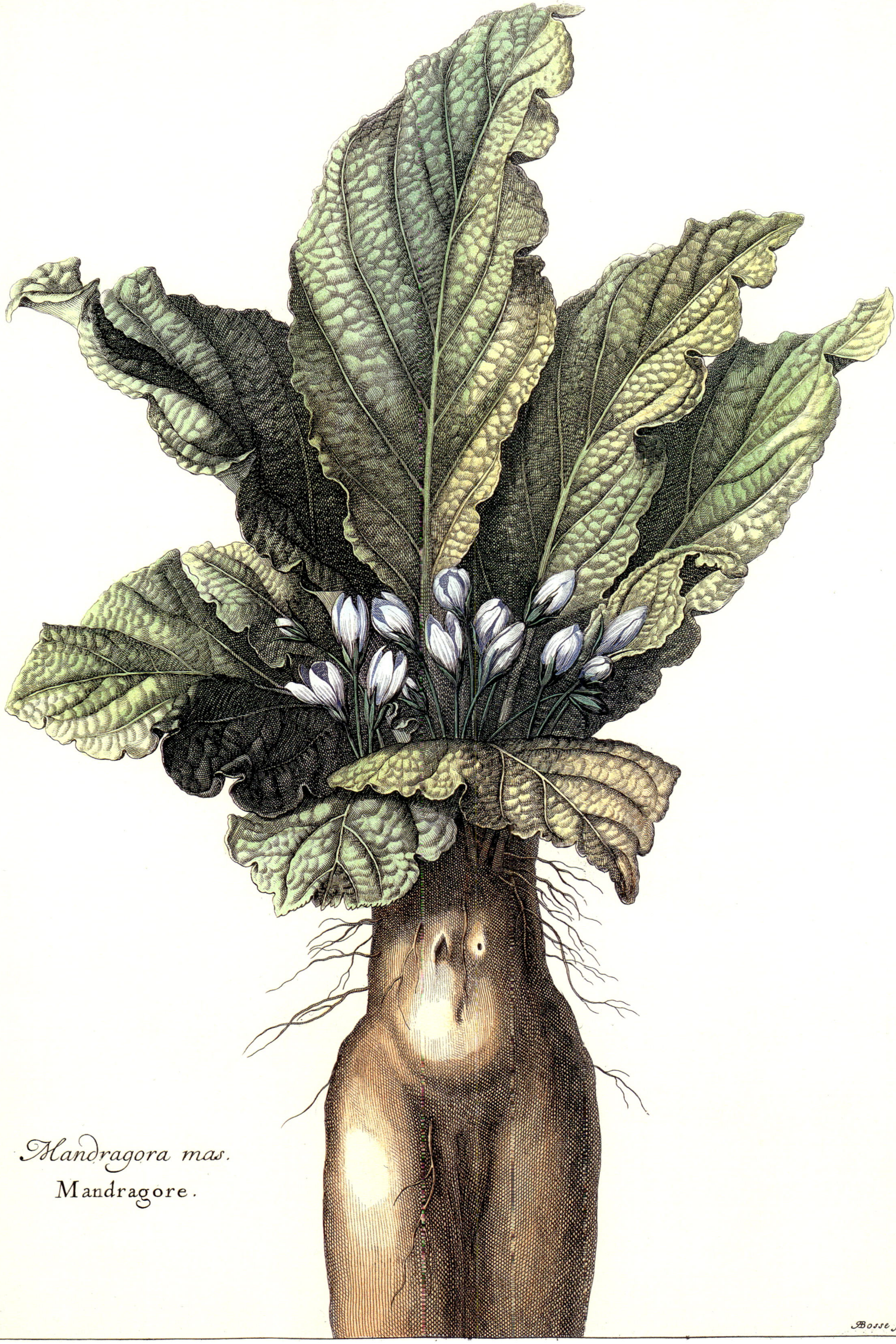

Mandragora mas.
Mandragore.

Bosse sculp.

Convolvulus cantabricus L.

CONVOLVULACEAE

The musicality of their name would seem to suggest that all *Convolvulus* twine around their supports like lianas, but this is far from the case. Some do not twine at all, including the plant named here; the same is true of *Convolvulus siculus* L, *C. lineatus* L., *C. lanuginosus* Desr. and *C. tricolor* L. Others twine to a limited extent: *C. arvensis* L., *C. althaeoides* L., and *C. elegantissimus* Miller. Finally there are those that twine in the way we normally associate with *Convolvulus* and which have now joined forces with ipomoeas and *Calystegia*. The Latin root *volvere*, which means to wind up, turn and even to 'spin around' has given rise in French to the words *vautrer* (to sprawl), *volute* (scroll) and ... *révolution*! *Volvere* itself contains the Indo-European root *wel*, to which the words *hélice* (helix) and *valse* (waltz) are related.

The best known members of the *Convolvulus* family are the field bindweed (*liseron des champs* in French) – or simply bindweed (*liseron*) – and the hedge bindweed, *Calystegia sepium* L. The former became known as *liseron* in French in 1538 because its flowers are similar to those of the lily (*lis*), but smaller. Although quite beautiful, these plants can be so invasive that, as with couch grass, people will go to any lengths to eradicate them. As eager consumers of nitrogen and organic matter, which gardens often contain in excess, they can completely colonise a garden in two to three years. And woe betide anyone who runs a rotovator over the area – this simply transforms the underground stems into cuttings that will go on to produce fine new plants the following spring. The only way to get rid of it is by persevering with a garden fork or 'garden claw'. And on top of all this, the plant is not edible as it contains toxic substances.

There are, however, some members of the *Convolvulus* family that have medicinal uses. These include the jalaps (which are ipomoeas), in particular the tuberous jalap whose Latin name, *Ipomoea purga* Hayne, could hardly be more descriptive. It was used medicinally in Mexico before being introduced into Europe by the first travelling botanists in the sixteenth century. Its tubers act as a drastic hydragogue purgative and are only used medicinally in cases of oedema.

The Cantabrian bindweed (the Cantabrian Mountains are the western extension of the Pyrenees), which is known in French as *liseron cantabrique* or *liseron-plante de Biscaye* (Biscay bindweed), grows in the south of France, mainly on rocky hillsides. A delicate and decorative plant, it flowers from May to July. Although in the past it was used as a vulnerary, to treat wounds by topical application, it has fallen out of use.

When cultivated as an ornamental in gardens in the south of France it produces superb pink flowers which are at their best when the sun first rises, with dewdrops highlighting the lines of hairs along the outside of the corolla.

Convolvulus cantabricus L. Engraving: Nicolas Robert
RMN, Louvre Chalcography Dept, Paris (modern print).

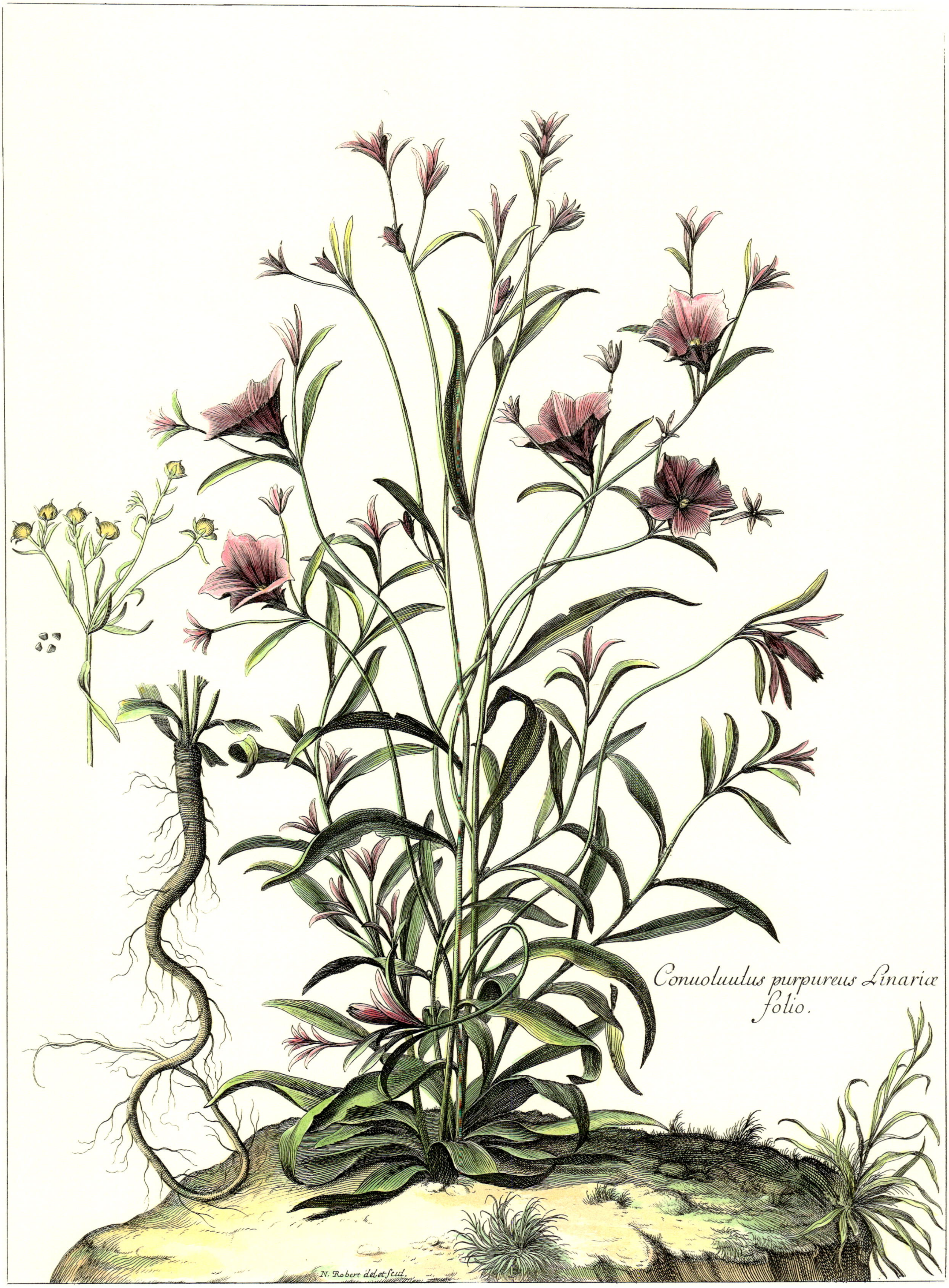
Conuoluulus purpureus Linariæ folio.
N. Robert del et Scul.

Convolvulus scammonia L.

CONVOLVULACEAE

Convolvulus scammonia L. is a hardy climber of the *Convolvulus* family, the best known member of which is the bindweed. The engraving shown here is a faithful representation of the morphology of this scammony, apart from the colour which is normally yellow. Does a blue variant of scammony in fact exist or should we assume, as has happened with other species, that the painter had allowed himself some licence in this respect? This plant does not grow in France. In Europe it is found in areas of scrubland and the dry, open forests of Greece, the Ukraine and the Crimea. It is also found in the Mediterranean basin, in Jordan, Lebanon, Israel, Sinai, Syria, Turkey etc.

At the time when it was thought that evil, and evil thoughts too, could be expunged through sweat, urine and excrement, scammony was much used as a purgative. Indeed, in one of his fables, La Fontaine describes how, when the tortoise appeared to have lost its reason, it was felt that it needed to be purged with hellebore seeds. The word 'purge' comes from the Latin *purgare* (to clean), derived from *purus* (without stain, without blemish). The Latin word *purus*, like the Greek word *katharos*, was used in religious language and by calling themselves 'Cathars' this religious group were claiming to be the 'pure ones'.

The concept of catharsis developed by Hippocrates (460–370 BC) advocated a balancing of the humours and hence involved relieving excess humour. Nerprun purgative (also called buckthorn) once formed part of an array of drastic purgatives, hence its evocative name: *Rhamnus catharticus* L. – RHAMNACEAE.

Daléchamps (1615) quotes Dioscorides (first century AD) on the subject of scammony: 'Its juice taken in the weight of a dram or four oboles ... purges the body downwards of phlegm and bile. To release the stomach, it suffices to take two oboles with sesame, or some other seed. For a stronger purge, three oboles of this juice are prescribed, two oboles of black hellebore and a dram of aloe.' He adds this entirely reasonable comment: 'With regard to what the Ancients have written on scammony juice, there are two things that merit consideration – which are how to identify what is good and the correct dose required for purging.'

There are four drugs known since ancient times for their purgative properties that are found in the following members of the *Convolvulus* family: *Ipomoea orizabenzis* Ledan (the Mexican scammony); *I. Purga* Hayne (the tuberous jalap); *Operculina turpethum* R. Br. (turpeth); and the scammony shown here, known as the Aleppo scammony. Among the active substances they contain are glucoresins, complex heterosides the genins of which are acid-alcohols associated with oses such as rhamnose and fucose. They form part of the group of hydragogue purgatives useful in oedema of cardiac origin (dropsy), which act by derivation through the inflow of water into the intestines. Resin from Aleppo scammony, which appears in the 1937 Codex and was used in the preparation of compound tincture of jalap and compound digitalis pills, has now been replaced by Mexican scammony.

In France, three common members of the *Convolvulus* family have been used as purgatives: field bindweed (*Convolulus arvensis* L.), hedge bindweed (*Calystegia sepium* (L.) R. Br.) and seashore false bindweed (*Calystegia soldanella* (L.) Roemer & Schultes).

Convolvulus scammonia L. Engraving: Nicolas Robert
RMN, Louvre Chalcography Dept, Paris (modern print).

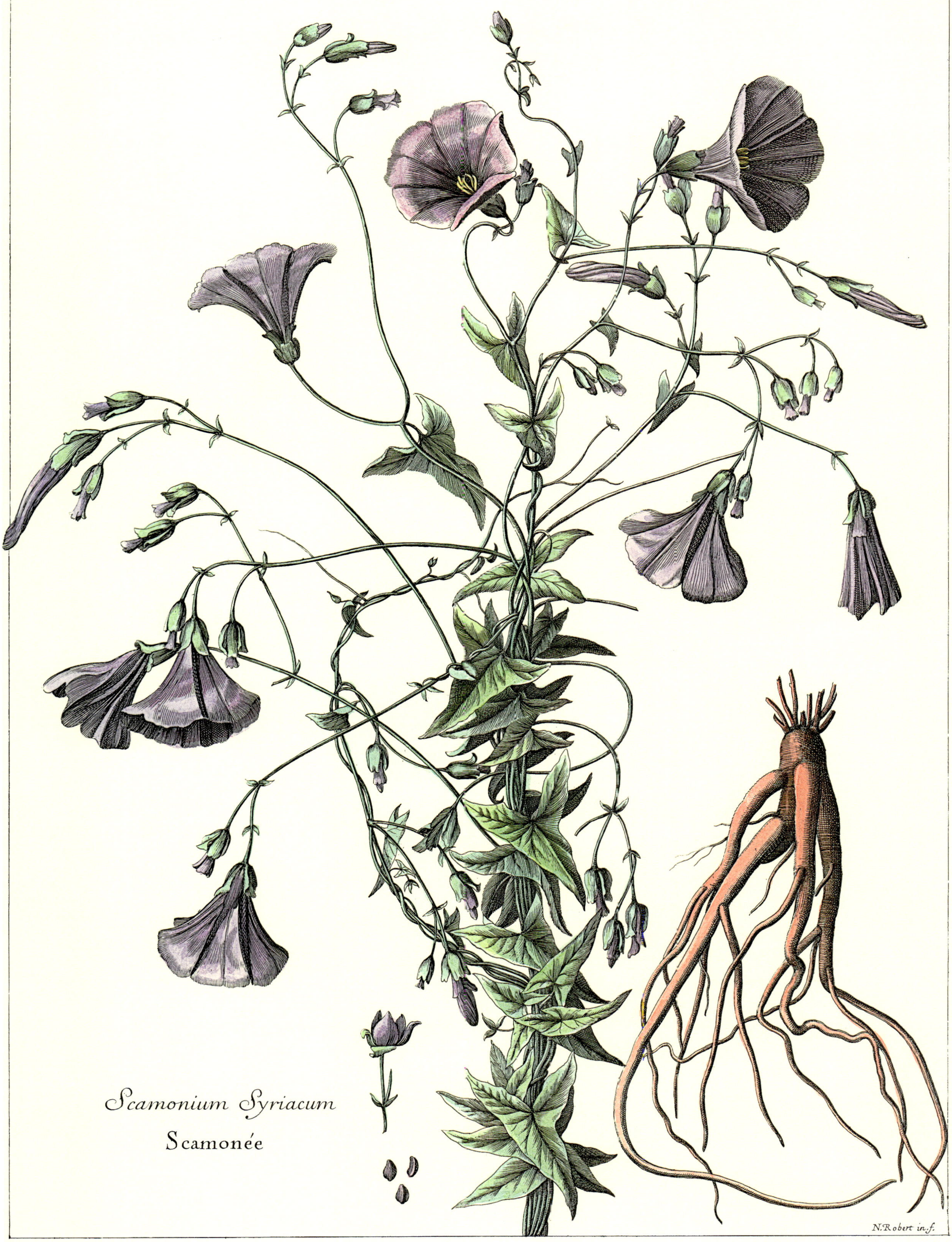

Scamonium Syriacum

Scamonée

N. Robert in. f.

Anchusa cretica Miller

BORAGINACEAE

This species does not exist in France. It is recorded in Italy (including Sicily), Greece, Albania, in the countries of the former Yugoslavia and in the south and west of the Balkan peninsula.

It is a very distinctive plant, its leaves covered in stiff, tuberous bristles and with pinkish-red buds that turn blue as they open – a common feature of the borage family and certain other blue-flowered genera. Though acid at first, the flowers become alkaline as they mature, changing colour in a similar way to litmus paper. In the past children in the Cevennes region, and probably in other areas too, would amuse themselves by getting ants to bite into borage or viper's bugloss flowers and watching how the acid left by the ant produced a small pink mark on the edge of the petal where it had been bitten.

In the borage family (as in others) a certain amount of reclassification has taken place: some *Lycopsis* have become *Anchusa* and *Nonea*; some *Anchusa* have become *Nonea* and *Pentaglottis*, some *Buglossoides* have become *Lithospermum*, and some *Lithospermum* have become *Lithodora* and *Amsinckia*, etc. Enough to confuse anyone, botanists included! The word 'bugloss' means 'ox-tongue', from the Greek *bous* (ox) and *glossa* (tongue). *Anchusa* comes from the Greek *anchousa* (make-up, paint) because the roots provide a red dye. There was once a plant known as *Anchusa tinctoria* L. but following name changes there are no longer any 'tinctorial' plants in the genus *Anchusa* – which is an aberration etymologically speaking.

The species is now called *Alkanna tinctoria* (L.) Tausch: dyer's bugloss. This plant is particularly happy growing in the sandy soil of the Dolomites, though it can be found right around the Mediterranean. Its roots are astonishingly red once the first layer has been removed. Women living on the shores of the Caspian Sea would use pieces of the root to redden their lips and cheeks. Alkannin, the colouring matter, is a naphtoquinonic pigment that is soluble in oils and other fats and was much used by Greek and Roman women. Dyer's bugloss was probably also used to dye everyday fabrics that were not long lasting, since no colorant based on this plant has ever been identified on ancient textiles. It did however enjoy a period of popularity during the vogue for Indian fabrics.

Another member of the borage family from the Mediterranean region, *Onosma echioides* L., known as the golden drop, also contains alkannin in its roots and is used in a similar way. There are records of recent use among the mountain dwellers of northern Greece. Along with ten or so other plants with purplish roots, it was used in *ratanjot*, one of the traditional dyes of Iran and India.

The borage family is a rich one and includes: the *myosotis* (forget-me-not) with its small, hairy leaves resembling the ears (*otos*) of a mouse (*muos*); comfrey, which assists the healing process; and borage itself which gives these plants their family name. Borage was used to such an extent as a sudorific that in France the name passed into popular use in the same way as aspirin has today in the expression *blanc comme un cachet d'aspirine* (white as an aspirin). In nineteenth-century France common expressions involving the notion of 'sweat' were sometimes replaced with *bourrache*, for example, *passer à la bourrache* meant to s'be taken to court' – a place where people did a lot of sweating! The word *bourrache* comes from the Arabic *abu araq* (father of sweat).

Anchusa cretica Miller. Engraving: Nicolas Robert
RMN, Louvre Chalcography Dept, Paris (modern print).

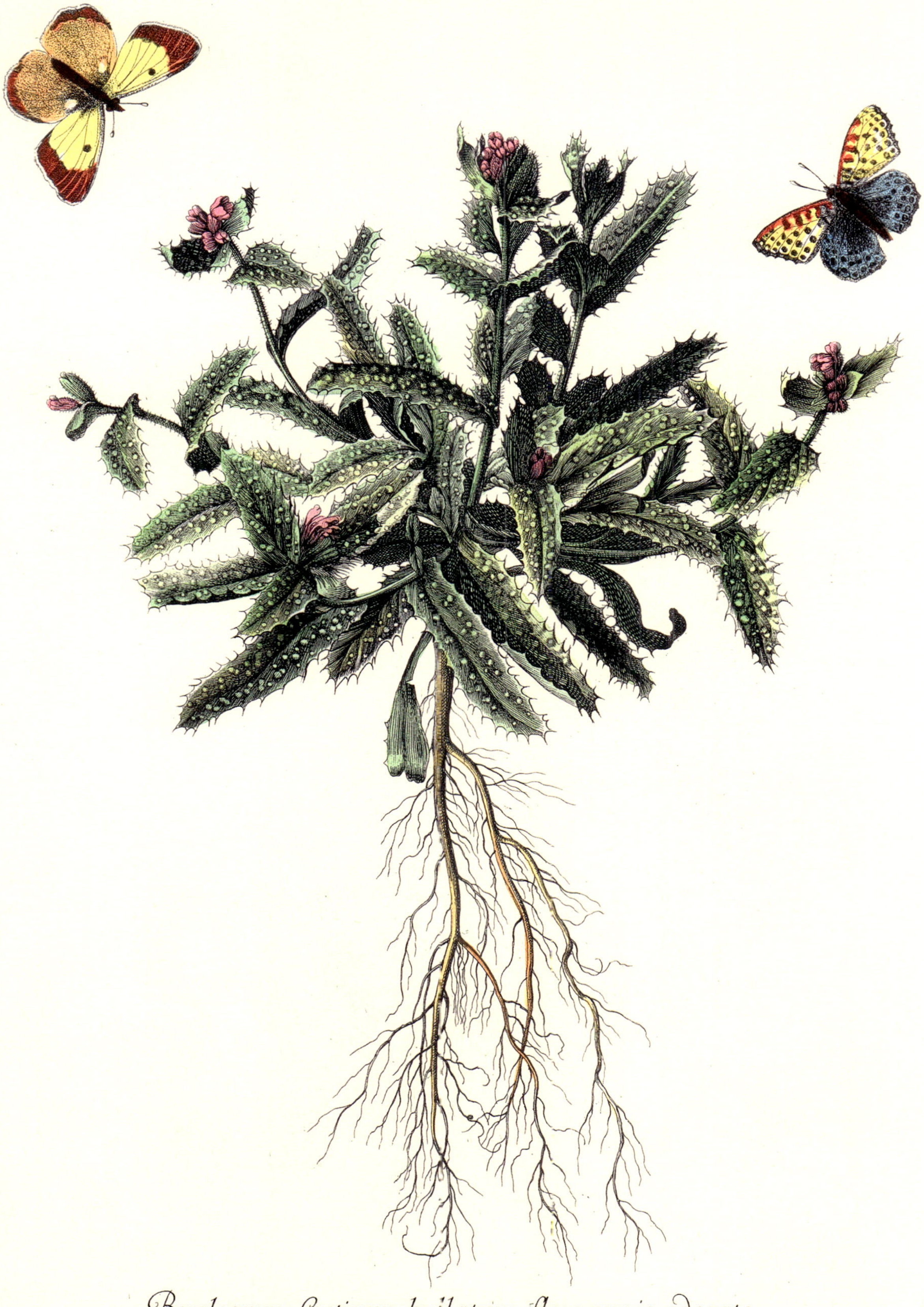

Buglossum Creticum bullatum, flore vario odorato.

Buglosse de Candie à feüilles bosselées, et à fleur odorante de plusieurs couleurs.

N. Robert del. et sculp.

Lomelosia graminifolia L.

DIPSACACEAE

Though the flower – or rather the inflorescence – of the scabious is quite familiar to amateur botanists, it poses more of a problem to the professionals for whom it is a point of honour to distinguish between *Scabiosa*, *Knautia* and *Succisa* and who more recently have also had to contend with *Lomelosia*, *Succisella*, *Sixalis* and *Pycnocomon*, not forgetting *Cephalaria*.

Traditional floras usually included the plant shown here as a scabious and it is entirely acceptable to call it the grass-leaved scabious. But before you can do that you first have to find it ... which is no easy task as it grows mainly in areas of scree and limestone rock in the Dauphiné Alps and in Provence.

The word 'scabious' comes from the Latin *scabies* (roughness) which in turn derives from *scabere* (to scratch), which is the origin of the word 'scabrous'. The plant was once used to treat scabies. Daléchamps (1615) mentions two types of scabious: 'Both are heat-giving, dessicant and soothing and are also detergent, for which reason they are excellent for evacuating excrement, and phlegm from the chest and the lungs ... They are also marvellously good against scabies, either by taking a decoction made from them for a number of days or by applying their juice either on its own or mixed with ointments to those suffering with scabies ... They are remarkable when applied as a liniment to buboes and it is even said that these are eradicated and disappear within three hours of application. Four ounces of their juice given with a dram of Theriac from the first day that a man is struck by the plague will be beneficial: but after this he should made to sweat in bed and to use this same medicine two or three more times. This juice is also excellent against snake bites ...'

More recently, the wide range of properties found in scabious, succisa and knautia has seen them used as tonics, astringents, vulneraries, digestives and depuratives.
In the early twentieth century, the herbalist Dr Leclerc achieved good results on the fluidification of bronchial secretions using succisa. This plant also appears to be quite effective when used topically: the leaves and roots are crushed and applied to areas of dermatosis and oozing eczema.

In the past, another teasel was much used in the carding of fine wools, so much so in fact that it was cultivated for this purpose. This was *Dipsacus sativus* (L.) Honckeny, known as the *chardon à foulon* and *chardon à drapier* in French and as fuller's teasel in English. The wild form *Dipsacus fullonum* L., (known in French as *cardère*, *cabaret des oiseaux* and *lavoir de Vénus*) was also used. The opposed leaves are joined at their base, forming a small bowl that collects rainwater and from which birds drink and in which many insects drown.

Lomelosia graminifolia L. Engraving : Nicolas Robert
RMN, Louvre Chalcography Dept, Paris (modern print).

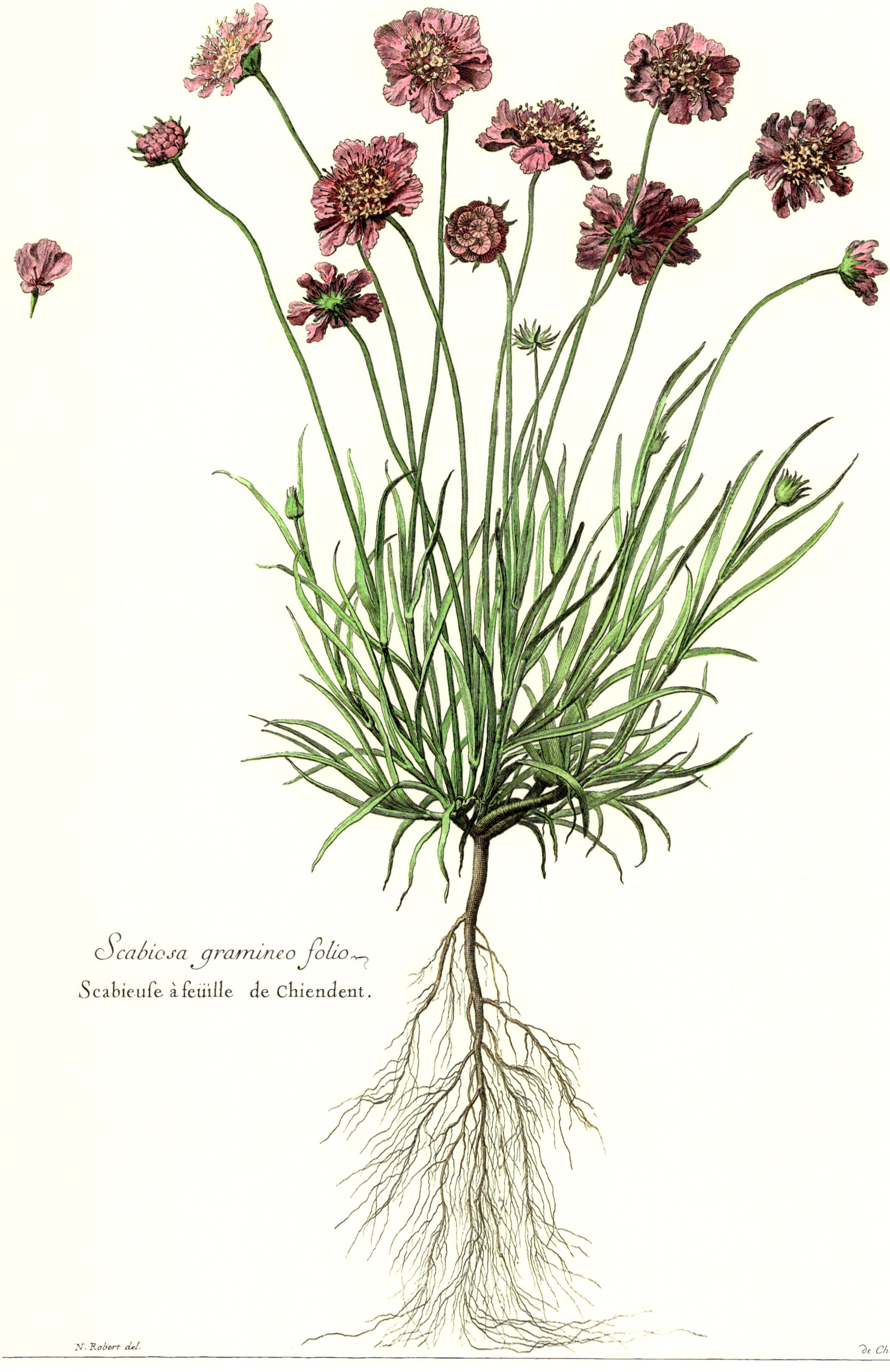

Scabiosa gramineo folio.
Scabieuse à feüille de Chiendent.

N. Robert del.

de Chastillon sculp.

Achillea distans Waldst. & Kit. ex-Willd. subsp. tanacetifolia Janchen

ASTERACEAE (EX-COMPOSITAE)

This is one of many species of achillea that form part of French flora, though it is not very common even in its preferred habitat, the high mountain areas of Savoy and the Hautes-Alpes. It is also found in mountainous areas of Switzerland, Italy, the Carpathians, the Balkans etc.

Its leaves are less serrated than those of the more common, plain-dwelling species *Achillea millefolium* L. (the common yarrow) but it possesses more or less the same properties.

In French the common yarrow is known by many names that reflect its popular use in treating wounds and bleeding: *herbe aux charpentiers* (carpenter's plant), *herbe à la coupure* (the cut plant), *saigne-nez* (nosebleed), *herbe aux militaires* (soldiers' plant) etc. Country people once applied it as a poultice of fresh, crushed leaves or as fresh juice to anal fissures, haemorrhoids, varicose veins and, wrapped in cloth, to wounds. The dried flowers of the plant were taken as an infusion to treat circulatory problems, painful periods, 'heavy legs' and disorders associated with the menopause. In fact, some people still plug their nose with crushed yarrow leaves when they have a nosebleed.

In addition to these uses, contemporary herbal medicine also recognises this plant as possessing choleretic and anti-inflammatory properties due to the presence of azulene and glucoprotein. However, ingestion of its essential oil is highly inadvisable as it contains thujone, a neurotoxic ketone.

It is a hardy plant and adapts well to dry soils. It is cultivated as a garden plant for its attractive corymbs of small, white and pink flowers. There are numerous varieties in a range of forms and colours, all equally elegant. It can also be finely chopped and sprinkled into a salad, a stew or even into mayonnaise.

The plant's finely serrated leaves give it its species name, while its genus name comes from the Greek hero Achilles who was an expert in medicinal plants, having been taught by the centaur Chiron. Achilles is credited with curing his friend Patroclus after the latter was wounded by an arrow when disembarking in Mysia.

Achillea distans Waldst. & Kit. ex-Willd. subsp. *tanacetifolia* Janchen. Engraving : Abraham Bosse
National Museum of Natural History, Paris.

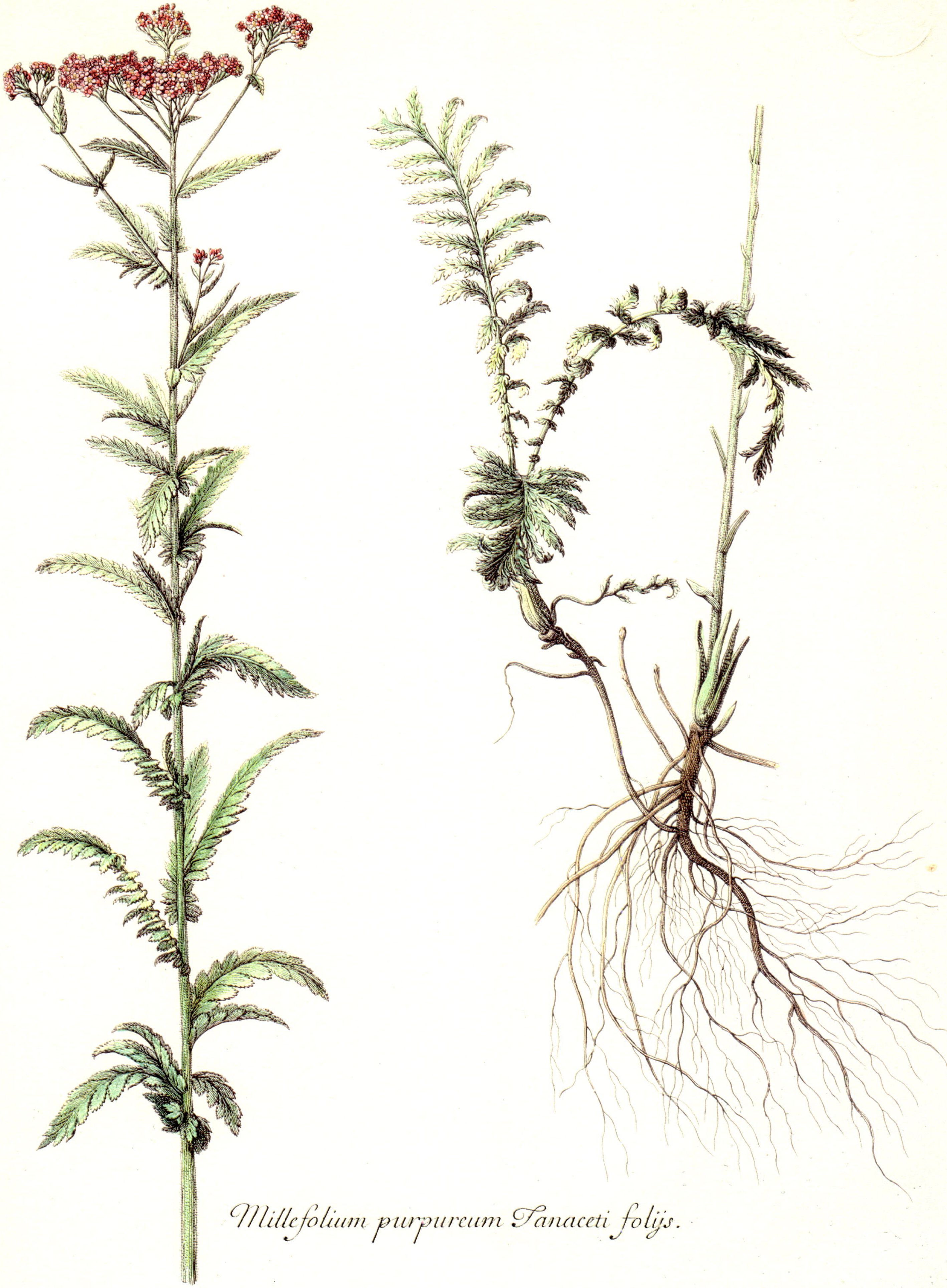

Millefolium purpureum Tanaceti folijs.

A. Bosse sculp.

Ambrosia artemisiifolia L.

ASTERACEAE (EX-COMPOSITAE)

While *Ambrosia maritima* may not be the ambrosia of the Olympian gods (a mere supposition challenged by some) it may well be the plant described by Dioscorides. In France only a few rare specimens are found on the coastal sands of the Bouches-du-Rhône, the Alpes-Maritimes and in Corsica, near Cervione. However, it is more common in other countries bordering the Mediterranean basin such as Spain, Italy, and Greece, and in the north of Africa – in places such as Nubia or Senegal.

It was once used as a medicinal plant for its stimulant, excitant and digestive properties. An aromatic, it was also used to flavour liqueurs for human consumption.

Recently, it has been found to have a strong molluscicide action (a rather uncommon term meaning that it kills molluscs) superior to that of *Ambrosia senegalensis* DC., and is particularly effective against the freshwater snail *Limnaea natalensis* which contaminates local water sources in Senegal, and against other snails that play host to parasites dangerous to man such as *Bilharzia*, *Fasciola hepatica* and *Schizostoma* (J. Belot, P. Bornarel and S. Geerts, 1986).

The engraving reproduced here in fact shows the common ragweed rather than the sea ragweed. The illustrator has taken particular care in the detail of the fruit, which are longer than they are broad in *A. artemisiifolia* and broader than they are long in *A. maritima*.

It would be good if one could find a beneficial use for *Ambrosia artemisiifolia* because, after colonising the north of America, it has invaded France and Europe, its pollen producing various respiratory allergies along the way. Indeed, it contains some powerful allergenic proteins, in particular *amb.a1* and *amb.a2*, along with profilin, a protein also present in celery, melon, bananas and sunflowers. According to Sylvie Déthiollaz, this would explain why 'people who are allergic to ragweed are also allergic to these plants' (from an article recently published by the Swiss Institute of Bioinformatics, in *Protéines à la une*).

The common ragweed is a prolific pollen producer – a single stem can release several million grains of pollen in a single day and they are so small that they are carried away by the wind. One gram of this pollen contains 90 million grains, each containing around fifteen allergens. When one considers that it requires only five grains of pollen per cubic metre to trigger rhinitis and that in the valleys of the Rhône and its tributaries, tens of millions of ragweed plants grow on the gravel banks along the edges of the rivers, allergy sufferers have every reason to be alarmed!

Originating in North America, the common ragweed was seen for the first time in Brandenburg, Germany, in 1863, and then in France in the Loire Valley in 1865. It grows abundantly in the administrative areas between the Rhône and the Vaucluse (affecting more than a hundred towns and villages in each administrative area). Only about twenty French *départements* have not yet reported the plant, which seems to be spreading inexorably. And this invasion is now affecting a much larger area of Europe.

In most countries, surveillance networks are being set up and allergy specialists, chemists, agronomists, ecologists and meteorologists are combining their efforts to combat what is a genuinely bad and unwelcome weed.

Ambrosia artemisiifolia L. Engraving : Nicolas Robert
National Museum of Natural History, Paris.

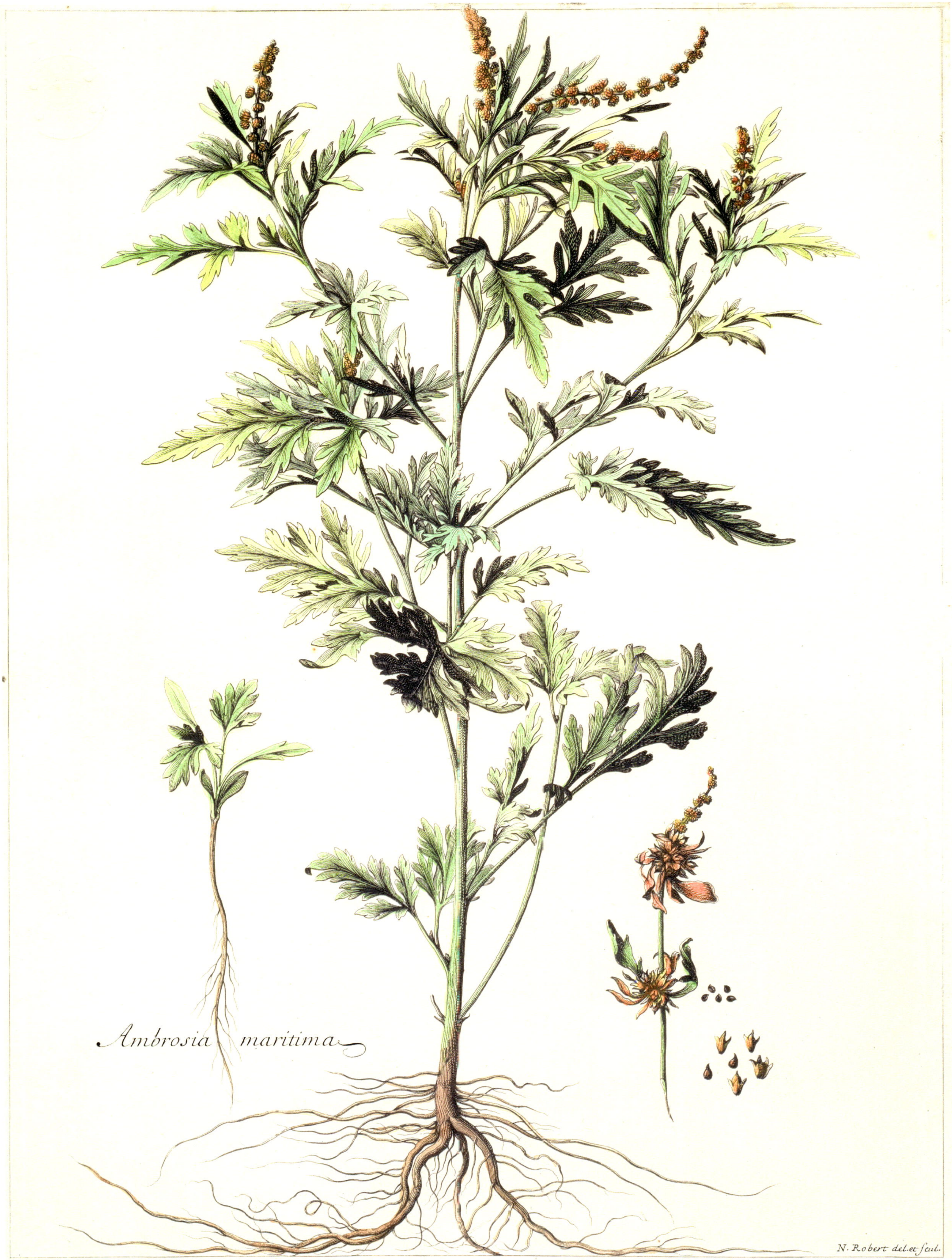
Ambrosia maritima
N. Robert del. et Sculp.

Aster alpinus L.

ASTERACEAE (EX-COMPOSITAE)

This aster, like many plants that grow at high altitudes, is not very large, however it can be spotted from a distance because of its beautiful flowerheads in contrasting colours, present from June to mid-September. Preferring sunny spots, it grows in mountain pastures and rocky areas. It is a hardy plant, well adapted to the rigours of the climate, with long, simple roots, a tough, woody stock and small, usually pubescent leaves. Interestingly, all mountain asters share the same colours: central yellow flowers and ligulate violet flowers.

The Alpine aster is considered rare in France, although it may be common locally. It is found at a height of between 800 and 2,800 metres (2,600 and 9,200 ft), in the Alps, the Pyrenees, the Jura, the Corbières mountains and the Cevennes mountains. Like other asters, it is visited by many insects, in particular bees and burnet moths.

There are two subspecies of *Aster alpinus*: *alpinus* and *cebennensis*. The latter is endemic to the high limestone plateaux of the Causses. In the valleys of the Causses there also grows a medium-sized aster with small capitula of yellow and violet flowers which is locally protected: *Aster sedifolius* L. subsp. *trinervis* (Pers.) Thell.

Another mountain aster, the Pyrenean aster (*Aster pyrenaeus* Desf. ex-DC.), which is taller and more vigorous, is a species endemic to France and Spain; it appears on the list of protected plants of France and also benefits from European protection status. Not only is it rare and extremely localised, it was also on the verge of disappearing because of botanists looking for unusual herbarium specimens and other rare-plant collectors who had no hesitation in uprooting specimens in order to plant them in some obscure lowland garden.

Fifteen species of *Aster* have been recorded in France. Of these, two are under threat of extinction – *Aster pyrenaeus* and *Aster amellus* L. – and two are classed as invasive plants – *Aster lanceolatus* Willd. and *Aster novi-belgii* L. It should be pointed out that the latter species are not native to France.

Aster alpinus L. Engraving: Nicolas Robert
RMN, Louvre Chalcography Dept, Paris (modern print).

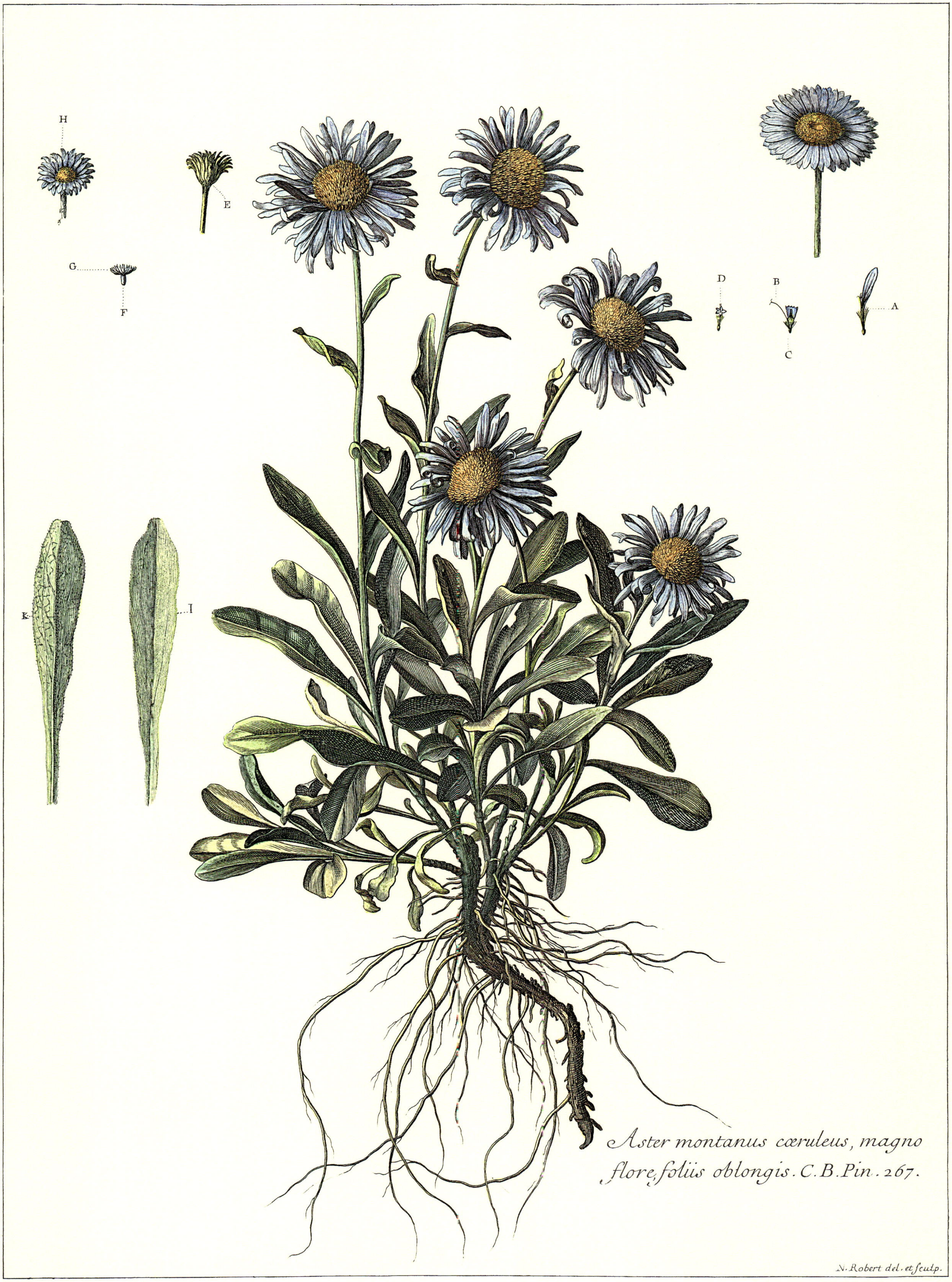

Aster montanus cœruleus, magno flore, foliis oblongis. C.B. Pin. 267.

Aster novi-belgii L., Aster novae-angliae L.

ASTERACEAE (EX-COMPOSITAE)

Eight species of *Aster* originating in North America are naturalised in France – *A. ericoides* L., *A. laevis* L., *A. lanceolatus* Willd., *A. novae-angliae* L., *A. novi-belgii* L., ssp. *laevigatus*, *A. novi-belgii* L. ssp. *Novi-belgii*, x *salignus* Willd. and *A. versicolor* Willd. – along with one originating in Central America and South America: *A. squamatus* (Spreng) Hieron.

The beauty of these plants and their hardiness have played a large part in their dissemination across the world, and horticulturalists continue to vie with each other in selecting, crossing and creating even more beautiful varieties.

It is difficult to be certain which species is illustrated here as the identification of American asters is based mainly on the morphology of the capitulum: whether or not the bracts are curved and to what extent the curve is outwards; and whether the inflorescences are glandulous, as in *Aster novae-angliae* or non-glandulous as in *Aster novi-belgii*. These two species, which are very similar in appearance, are the ones that most closely resemble the specimen shown here. However, the colour used on this engraving does not correspond to any of the original asters and the plant is more reminiscent of a cultivar. In view of this uncertainty, it is better to look at both species.

Aster novi-belgii, an autumn-flowering species, comes from an area that was known as 'New Belgium', that is to say New Holland, the first seal of which, dating from 1623, bears the image of a beaver and the words '*Sigillum Novi Belgii*'. The area is now New York State, including its capital which was once known as New Amsterdam. In March 1664, King Charles II of England granted New Amsterdam to his brother James II, Duke of York, which is how it came to be called New York.

In 1710, it was the turn of this plant to set out to colonise Europe and it has been so successful in this that it is now listed as an invasive species in *Plantes invasives en France* (Muller, 2004), along with *Aster lanceolatus*. It colonises more or less the same habitats as *Aster novae-angliae*, that is to say the edges of rivers and streams, open, damp forest interiors and areas of wasteland. It is in damp situations that these two asters prove to be most invasive, posing a threat to vulnerable indigenous species. Apart from this drawback, it is a superb plant and its many cultivars are attractive additions to the garden.

Aster novae-angliae, as its name suggests, comes from the extreme north-east of the United States, from an area once known as New England. It is found growing wild in the dry fields of neighbouring Quebec where it is also known as the 'New England aster'. Introduced into Europe at the same period as *Aster novi-belgii*, this autumn-flowering species spread rapidly. An unusual feature of its glandulous capitula is that they exude an odour somewhere between camphor and turpentine which is even stronger when they are crushed. Many cultivars in different forms and colours are sold commercially.

The plant's rhizomes have been used for their astringent and febrifuge properties. In Japan, the shoots and young leaves are picked before flowering. They are then cooked and steeped in water to remove excess bitterness and are eaten as a vegetable or as a soup, seasoned with *gomasio*, *soyu* or *miso*. In China, it is the young shoots of *Aster trinervius* Roxb. that are eaten. Known as *Man lan* or *Tse kiu*, they are prepared in the same way.

Aster novi-belgii L., *Aster novae-angliae* L. Engraving: Louis de Chastillon
RMN, Louvre Chalcography Dept, Paris (modern print).

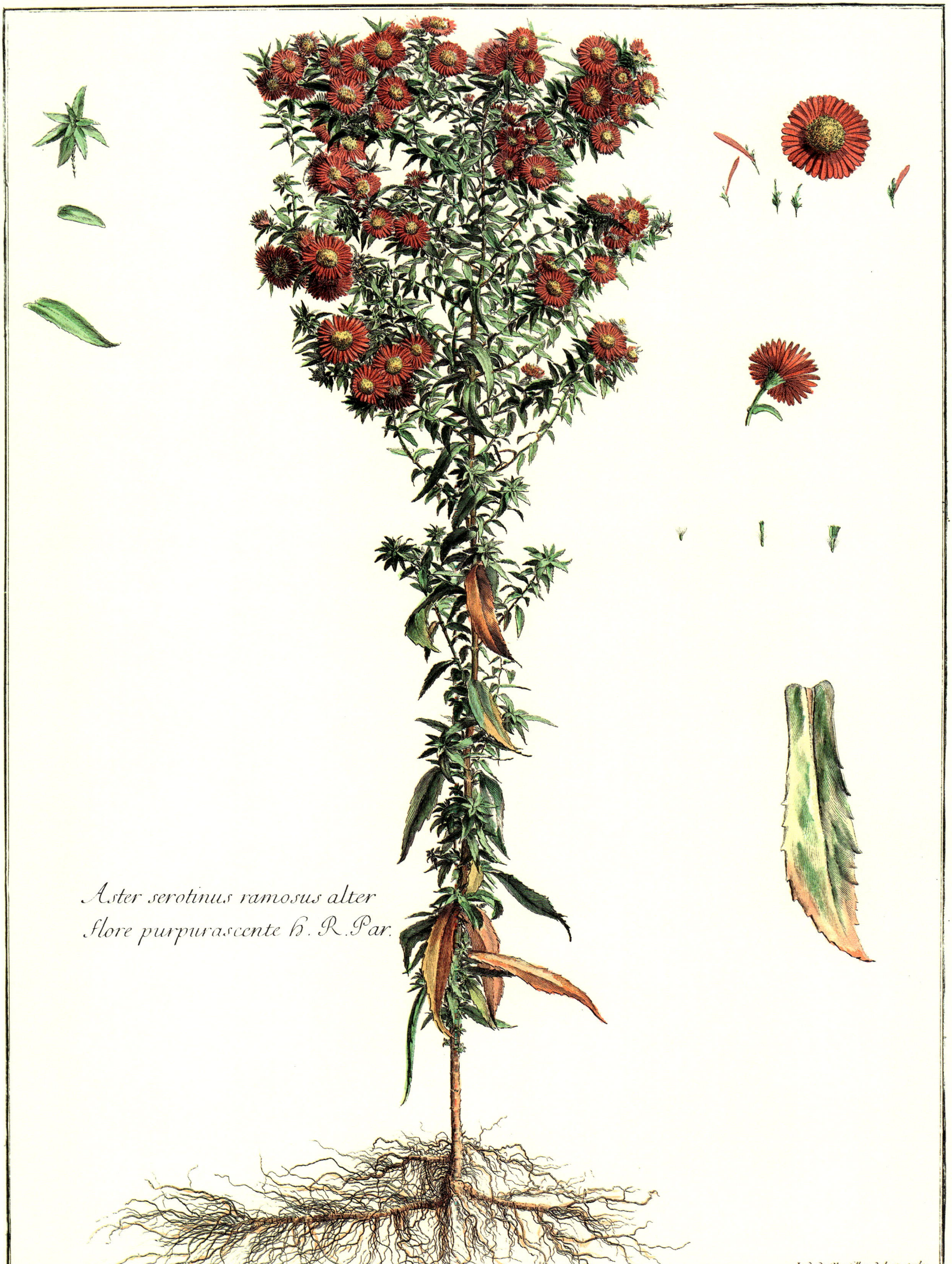

Aster serotinus ramosus alter flore purpurascente h. R. Par.

Lud. de Chastillon del. et sculp.

Aster amellus L.

ASTERACEAE (EX-COMPOSITAE)

This very elegant plant is under threat due to the abandonment of hay meadows and grazing land and the encroachment of building on its natural habitat. It grows on the most exposed limestone hillsides in eastern France, and in dry meadows and sunny clearings from the Champagne and Lorraine regions to the eastern Pyrenees. It is also found in central Europe and western Asia. In France it now benefits from protected plant status under the ministerial order of 20 January 1982, amended by the order of 31 August 1995.

In French it is known as the *marguerite de la Saint-Michel* (and in English as the Michaelmas Daisy) after the feast of St Michael, which is celebrated on 29 September. *Aster amellus* is a pubescent plant, growing to 40 to 50 cm (16 to 20 in) in height, with radical, oblong, lanceolate leaves tapering into the petiole, while the leaves of the flower stalk are slender, sessile and pointed. A hardy plant, it flowers from August to October, depending on the situation, producing splendid capitula with central yellow flowers surrounded by ligulate flowers that are generally bluish-mauve. A melliferous plant, it is much visited by bees who are attracted to it by the light vanilla odour it emits at the hottest part of the day and in the evening.

Aster amellus, also known in French as *œil du Christ* (Christ's eye) has been used to produce numerous cultivars. Garden plant catalogues offer an impressive list of varieties including 'Rudolph Goethe', 'Madame riche', 'Hermann Löns', 'Mira', 'Sonia', 'King George', 'Nocturne', 'September Glory', 'Coombe Fishacre' and 'Violet queen'.

The word *Aster* is Latin for star, and according to certain etymologies, *Amellus* is related to the (male and female) Arabic forenames *Amal* and *Amel*, which mean 'hope' – an entirely fitting coincidence for this endangered plant.

Aster amellus L. Engraving: Nicolas Robert
RMN, Louvre Chalcography Dept, Paris (modern print).

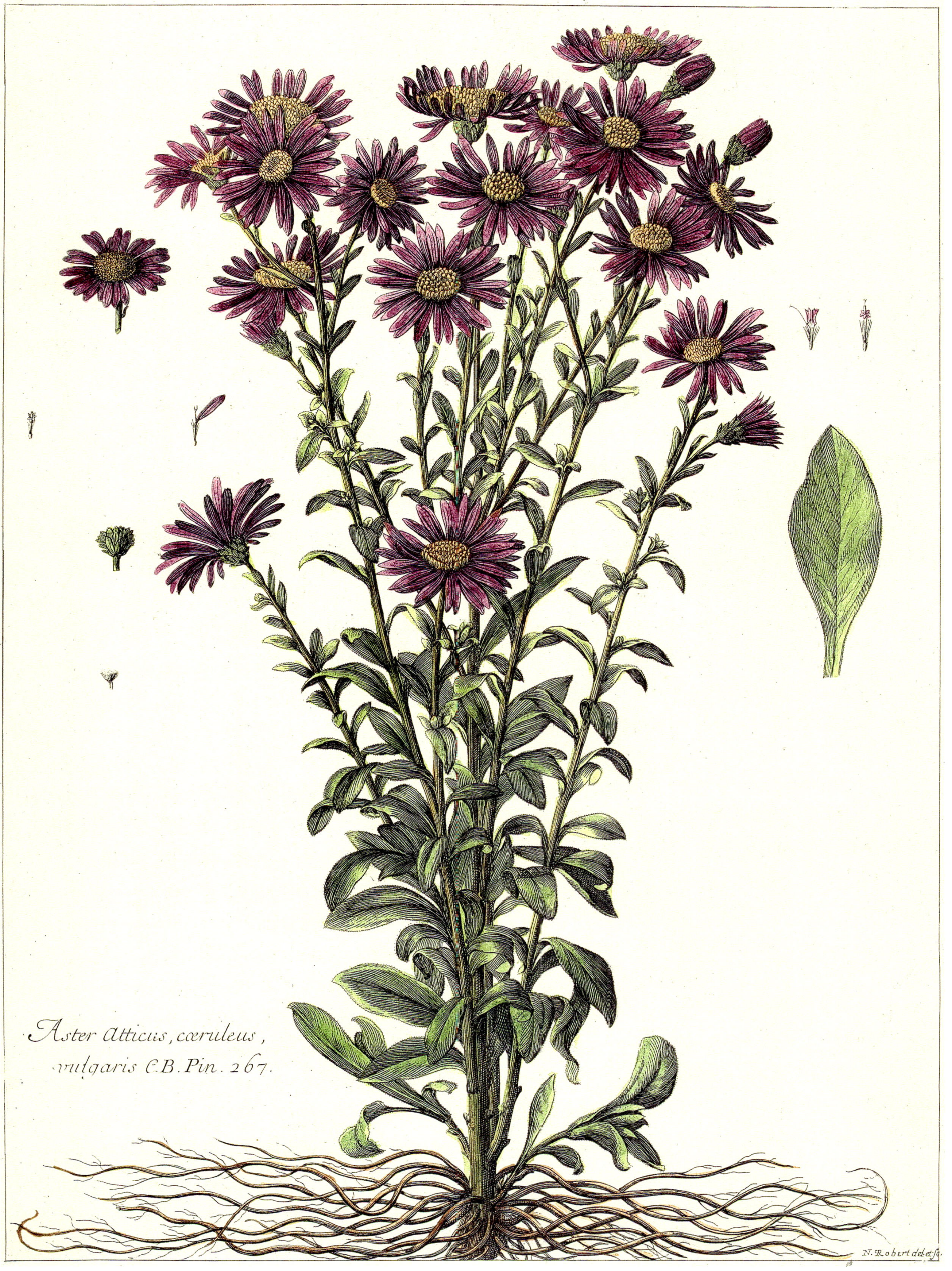
Aster Atticus, cœruleus, vulgaris C.B. Pin. 267.
N. Robert del. et sc.

Grindelia squarrosa (Pursh) Dunal

ASTERACEAE (EX-COMPOSITAE)

The country of Latvia, which is about the same size as Switzerland, is divided into twenty-six cantons, with Riga as its capital. It was here that David Hieronymus Grindel (1776–1836), who would go on to become the pride of his native city, was born. Grindel studied in Jena and then worked as a pharmacist in Riga. He went on to teach chemistry and pharmacy at Dorpat University, where he became rector. At the age of forty-four he began studying to become a doctor. Many years earlier, in 1803, he had founded the Riga Pharmaceutical Society and published Russia's first pharmaceutical journal, known as the *Russisches Jarhbuch der Pharmazie*, containing articles on pharmacy, chemistry and botany. Grindel also studied the flora of the Baltic Sea area. Karl Ludwig Willdenow (1765–1812), a German botanist, pharmacist and doctor (whose name is better known in the abbreviated form 'Willd.') named the genus *Grindelia* for Grindel. Willdenow was director of the Berlin Botanical Garden from 1801 to 1812 and made studies of many South American plants brought back by the explorer Alexander von Humboldt. He left behind a herbarium of over 20,000 species.

Grindelia, most of which originate in North and Central America, can be identified from the outward curving bracts of their capitula. The plant contains a sticky, gummy substance containing diterpenic acids, hence the English name 'gumweed'. A number of species are prized for their antitussive and antibacterial medicinal properties, principally *Grindelia robusta* Nutt., but also *G. squarrosa* and *G. camporum* Green. *G. squarrosa* is a large plant, at least 1 metre (over 3 ft) tall and branching towards the top. Cultivated as a medicinal plant, like the species *robusta*, it is now naturalised in vast colonies in the Ukraine. In Europe it is found only in the east and west of Russia.

The phenol acids and flavonoids, concentrated mainly in the flower heads, exert an antispasmodic action by depressing the bulbar centre of the cough. There is an additional expectorant effect due to the presence of saponosides. Grindelic acid is effective in combating the whooping cough bacillus (*Bordetulla pertussis*).

In rural areas, particularly in the Cevennes, many elderly people recount how as children they were saved from whooping cough by snails. In the early twentieth century, few or no remedies existed for certain illnesses and whooping cough often proved fatal. This remedy involved placing snails in a dish, sprinkling them with sugar to make them froth and then getting infected children to drink this sugary foam. In fact, snail froth contains a substance that is genuinely antitussive and coughs are still treated with *helix*-based syrups.

Grindelia squarrosa (Pursh) Dunal. Engraving : Nicolas Robert
National Museum of Natural History, Paris.

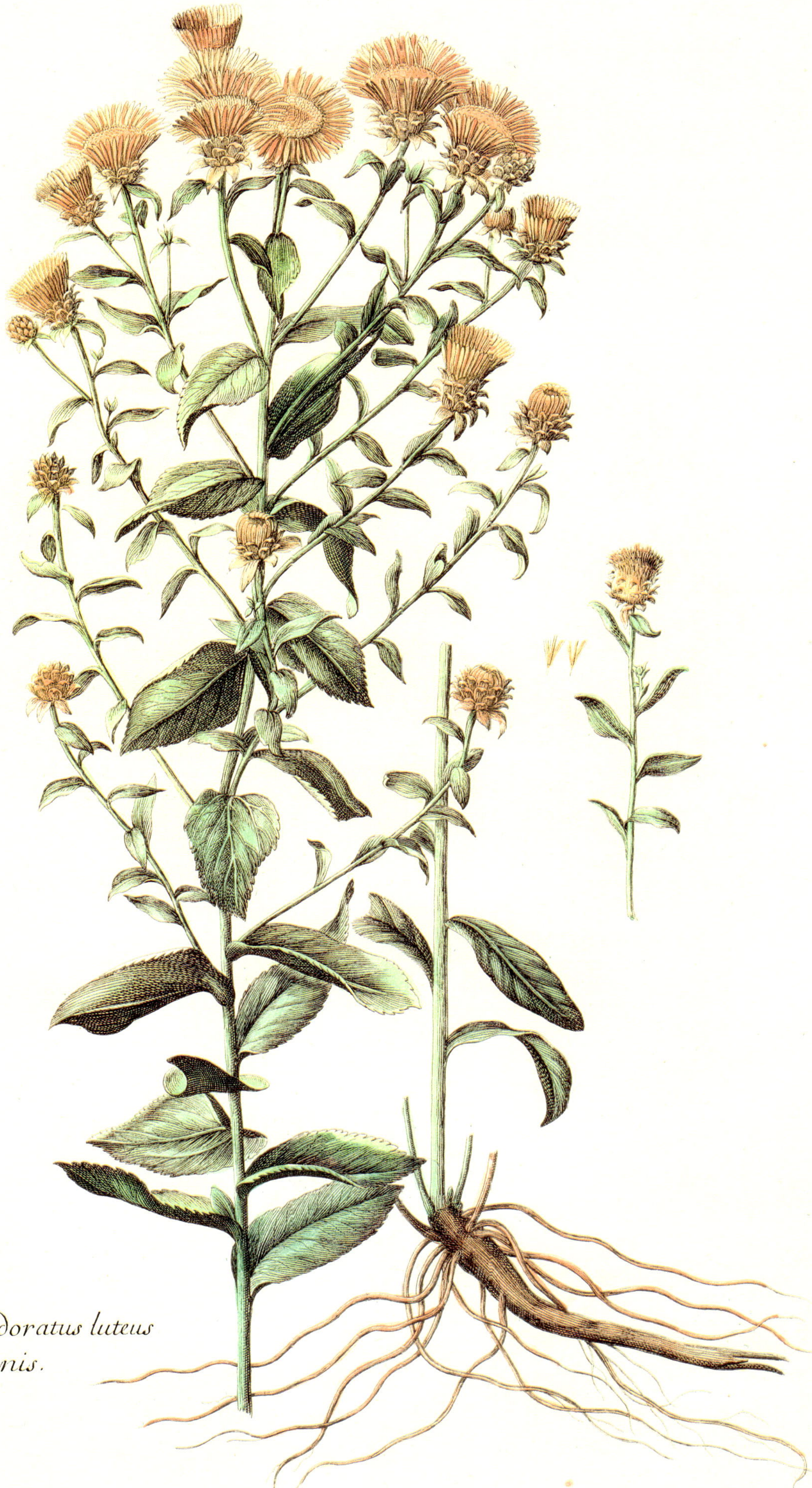

Aster conizoides odoratus luteus perennis.

N. Robert in. f.

Anthemis tinctoria L.

ASTERACEAE (EX-COMPOSITAE)

Some fine readjustments have taken place in deciding which species to include or exclude from the genus *Anthemis*. This has meant that some *Anthemis* have now become *Chamaemelum*, the best known of which, *nobile*, is the Roman chamomile, while some *Cota* have now joined the genus *Anthemis*, including one formerly known as *Cota tinctoria* Gay, now renamed *Anthemis tinctoria* L., which has two subspecies. As for *Buphtalum*, only one species of this genus is now found in France, the *salicifolium* (willow-leaved), the others having been moved to the genera *Astericus*, *Nauplius*, *Pallenis* and *Telekia*.

Buphtalum means 'ox eye', from the Greek *bous* (ox) and *ophtalmos* (eye) – no doubt an allusion to its slightly convex capitulum. Indeed, the term has been retained in French to refer to yellow or dyer's chamomile (*œil-de-bœuf*), as well as in Italian (*occhio bovino*) and in English (yellow oxeye).

This plant flowers in June and July. It is found in scattered locations on stony hillsides, preferably limestone, in areas up to a height of 1,600 metres (5,250 ft), right along the eastern border of France, from Lorraine to the Mediterranean and from the Alpes-Maritimes to Hérault. It has also been reported in several cantons of Switzerland.

In the past, the yellow oxeye (or golden marguerite) was applied as a vulnerary to wounds and taken as an infusion for its aperitive and digestive properties. However, it was principally used and known as a source of dye. Its colorants are the flavone heterosides (from the Latin *flavus* meaning yellow) apigenol and luteolol, and pigments from the flavonol group (quercetagetin and patuletin, which are also found in French marigolds and African marigolds: *Tagetes*). Other plants with yellow colouring flavonoids include weld, black oak, saw-wort, buckthorns and dyer's broom.

After applying a mordant to the wool fibre or cotton fabric to be dyed – the mordant encourages the dye to take and bind and can also change its colour through a chemical reaction – the material is plunged into a decoction of fresh, macerated *Anthemis* leaves and flowers. This is simmered together for a certain period – a few minutes or longer depending on the species – stirring to ensure that the dye is evenly distributed. The fabric or skeins of wool are then removed and left to dry in the shade, either out of doors or in a well-ventilated room. To avoid wool 'matting', it must be plunged into a vessel of very hot water and left to cool gradually.

Using potash alum as a mordant will produce a clear to lemon yellow; using a tin mordant (stannous chloride) will give a golden yellow; and copper sulphate will give an olive green.

Among the many stories associated with dye-producing plants, three plants (weld, goldenrod and buckthorn, with its so-called *graines d'Avignon* or 'Persian berries') are associated with the history of the Jewish diaspora. In 849, the caliph al-Mutawakkil ordered Jews and Christians in Persia to wear the yellow cloak and hood of the Persians. To avoid unions between Christians and non-Christians, at the Lateran Council of 1215, Pope Innocent III ordered Jews and Saracens living in Christian provinces to wear a yellow oblong disc or a yellow hood. In 1269, Louis IX, went further still and obliged them to wear two yellow identifiers, one on the chest and one on the back. It was another five centuries before this shocking practice was abolished (1791), but barely 148 years later it reappeared in Germany, on 1 December 1939 with an armband, then on 19 September 1941 with the yellow star, and in France on 29 May 1942.

Anthemis tinctoria L. Engraving : Abraham Bosse
National Museum of Natural History, Paris.

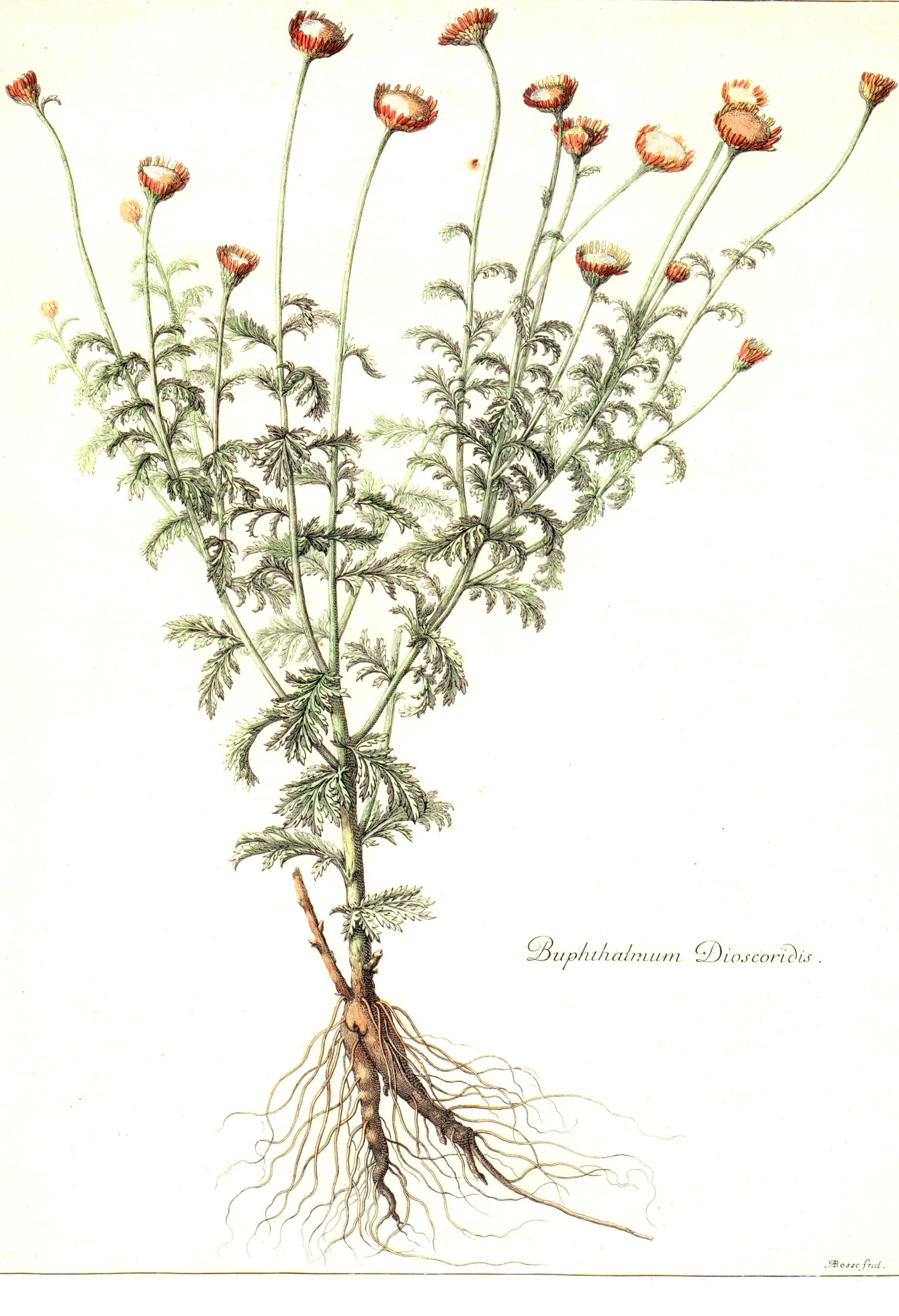

Buphthalmum Dioscoridis.

A. Bosse fecit.

Centaurea africana Lam.

ASTERACEAE (EX-COMPOSITAE)

The plant shown here appears to be *Centaurea africana*, a plant that prefers dry, open woodland and shrubland, and is found in Spain, Portugal and North Africa. If the illustrator has been scrupulous in his presentation, there is a slight doubt about the identity of the plant: the Alpine knapweed has the same capitulum but more pinnatisect leaves. In France the latter species is only found in a small valley near Sospel and is therefore listed in the *Livre rouge de la flore menacée de France* (1995) as an endangered species.

The composites are a vast plant family, including groups that are so different in terms of inflorescence that it has been divided into three sub-families: the corymbiferae (or radiate forms) such as daisies and marigolds; the cynarocephalae, such as the thistle and knapweed; and the liguliflora such as the golden thistle, salsify, and dandelion.

Centaurea are easily identifiable from their globular involucrum, their leaflets tipped by membranous, sometimes thorny, ciliate or feathery appendages, and their smooth akenes (fruit), which may be topped by a small tuft of hairs, known as the 'pappus'.

Few centaurea have medicinal uses, the exceptions being the cornflower and, in the past, the purple starthistle and the brown knapweed. The inflorescences of the common cornflower (*Centaurea cyanus* L.) contain polyines and anthocyanosides which are useful in the treatment of eye inflammation and as a complementary treatment for rheumatic conditions. The purple starthistle (*Centaurea calcitrapa* L.) and the brown knapweed (*Centaurea jacea* L.) were once used medicinally as aperitives, vulneraries and febrifuges.

The leaflets of the purple starthistle's involucrum terminate in a strong, canaliculate thorn, as formidable as the slender, pointed spines of the yellow starthistle (*Centaurea solstitialis* L.). The term *chausse-trappe* (the French name for the purple starthistle) refers to a small device known as a tribulus, with four iron spikes arranged in such a way that when placed on the ground one point would always remain upright. It was a type of weapon used to halt cavalry charges and wound infantrymen in the feet. The word comes from the old French *canketrepe* (1180), then *chauchetrepe* (1220), which comes from *chauche* (to crush, press) – from *calcare*, which is related to *cauchemar* (nightmare) – and *trepe* (to walk on, jump), which is the source of *trépigner* (to stamp).

With resolute optimism, Daléchamps (1615) describes some of the medical uses to which brown knapweed was put at that period: 'The powder of the great cornflower is good for those who have fallen from a height, or who are bringing up blood, if they are given it with plantain water, horsetail water, or common comfrey water, and for general healing. Its juice applied as a liniment heals fresh wounds and cures mouth ulcers. The leaves if cooked in wine and taken as a beverage are good for those who have been stung by scorpions and harvestmen, in fact they are good against all kinds of poisons.'

The word 'centaurea' is derived from *Kentauros*. The centaurs were mythological creatures who were half-man, half-horse. Chiron was the most famous, wisest and most knowledgeable of these fabulous creatures and was the fruit of the union between Cronos, who had taken the form of a horse, and the nymph Phylira. An expert on medicinal plants, Chiron is said to have discovered the virtues of the centaurea and the plant was therefore named in his honour. He also trained Achilles, Jason and Asclepios. Asclepios was a hero and god of medicine and had a number of children: two sons, Machaon the surgeon and Podaleirios the practitioner; and three daughters Hygeia, health; Iaso, healing; and Panacea, the universal plant remedy (from the Greek *pan* meaning 'all' and *akeios* meaning 'plant remedy'). Panacea, personified by the goddess Panakeia, came to mean an imaginary plant that could be used to cure all ills.

Centaurea africana Lam. Engraving: Nicolas Robert
RMN, Louvre Chalcography Dept, Paris (modern print).

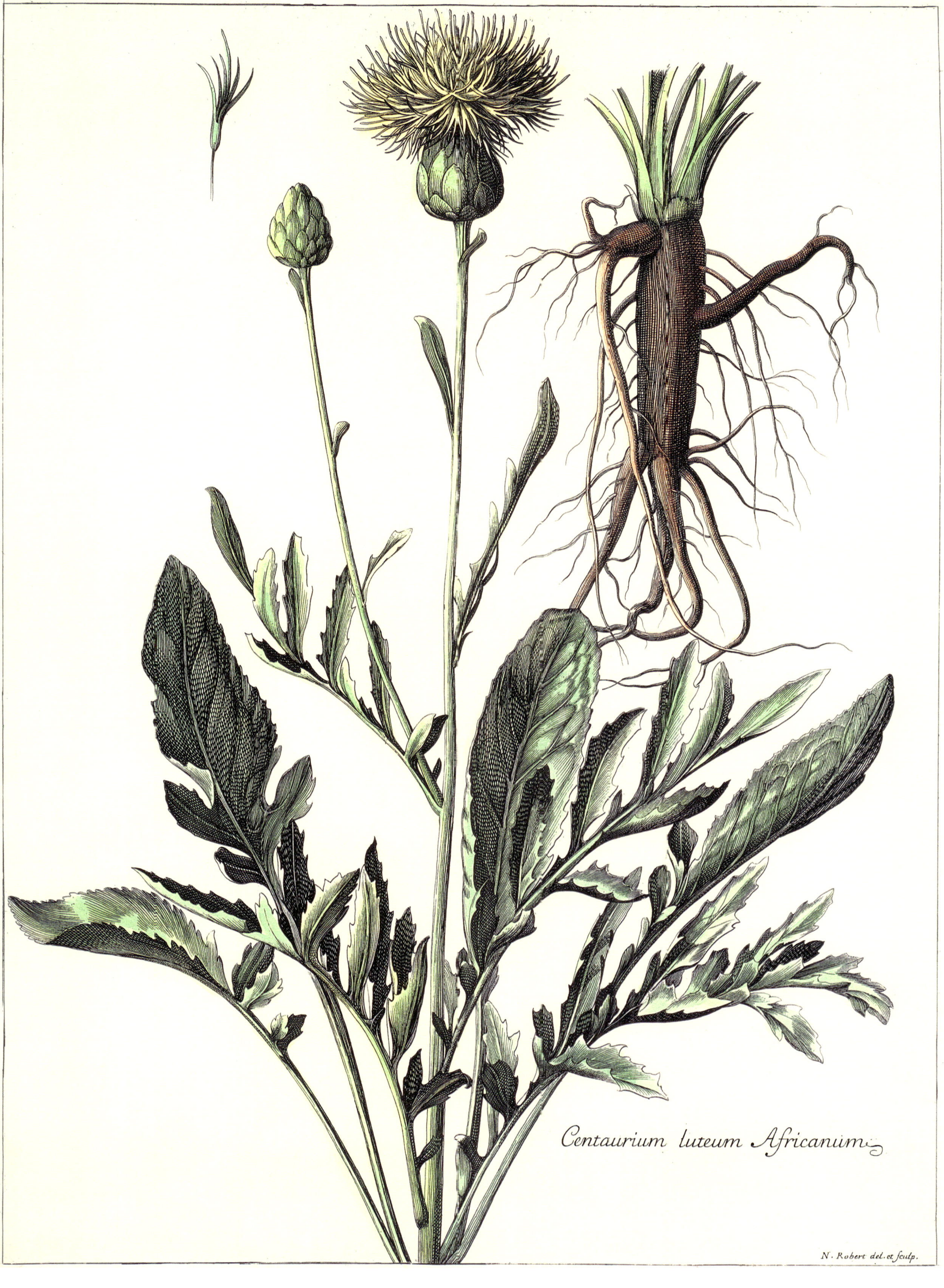

Centaurium luteum Africanum

N. Robert del. et sculp.

Helianthus annuus L.

ASTERACEAE (EX-COMPOSITAE)

It would have been difficult to call this plant with its large yellow capitulum surrounded by radiating ligulate flowers anything other than 'sun flower'. The botanical name comes from the Greek *helios* (sun) and *anthos* (flower). *Annuus* indicates that the plant is an annual. We have all at some time stood in wonder at the sight of a sunny field of these flowers. In fact the capitula can contain as many as 2,000 flowers, depending on the variety, and these will produce fruits that are generally referred to as 'seeds'. The highly decorative ligulate flowers are in fact sterile, whereas the central flowers or florets, arranged in a double spiral, are fertile.

Another species, *Helianthus tuberosus* L., is also well known: its tuberous rhizomes were eaten in such large quantities during the Second World War that many elderly people prefer not to hear them mentioned. This species is in fact the Jerusalem artichoke (*topinambour* in French), which has a delicious artichoke flavour and which is gradually coming back into vogue in kitchen gardens. It originally grew wild in the south of Canada and the north of America prior to the period of colonisation. In Canada it was eaten by the Hurons and Algonquins, who introduced it to the French colonists. It had reached France by 1610, where it was known as the 'Canada truffle' or 'Canada apple' and its cultivation soon spread. In 1613, Isaac de Razilly brought six members of the Brazilian Tupinambas tribe to the French court and, somewhat strangely, the name *topinambour* replaced the Canadian names the plant had been given.

The wild sunflower had already been eaten for many centuries. Introduced into Europe in 1596, for many years it was cultivated as an ornamental. People at that period were very impressed by its inflorescence. This is what Daléchamps has to say about it in 1615: 'This plant, which is known as the *Peru chrysanthemum* because it is said to have been discovered in Peru and in certain other provinces of America, from whence it was brought here, surpasses all other plants that produce golden flowers, both in beauty and in size. Some call it *Planta maxima*, others *Flos Solis* or *Sol Indianus*; the Portuguese call it *Gigante*. It is a marvellously tall plant, there being in Spain some that are twenty-four feet in height ... When the flower is fully open, it resembles a plate, as much as a foot across, yellow to begin with, then turning a reddish brown. The entire flower leans heavily towards the ground. Then there is another sun flower which is smaller and produces a branching stem. Its leaves are similar but its flower is smaller and the leaves are arranged around it in the form of a star.'

From this account it would seem that the plant was very quickly selected and bred, as the inflorescences of the original sunflowers were more like those of the Jerusalem artichoke. The sunflower was not cultivated for its oil until the middle of the twentieth century. The practice started in Russia where the Orthodox clergy were particularly strict about Lenten fasting. The Russians hit on the idea of extracting sunflower oil for consumption as it did not appear on the list of forbidden foods. In other European countries, sunflower cultivation did not really become common until the 1950s. The unsaturated fatty acids found in virgin sunflower oil give it very useful cholesterol lowering properties.

However, one question still remains: does the sunflower really move with the sun? Opinions are divided, even among experts.

Helianthus annuus L. Engraving: Abraham Bosse
RMN, Louvre Chalcography Dept, Paris (modern print).

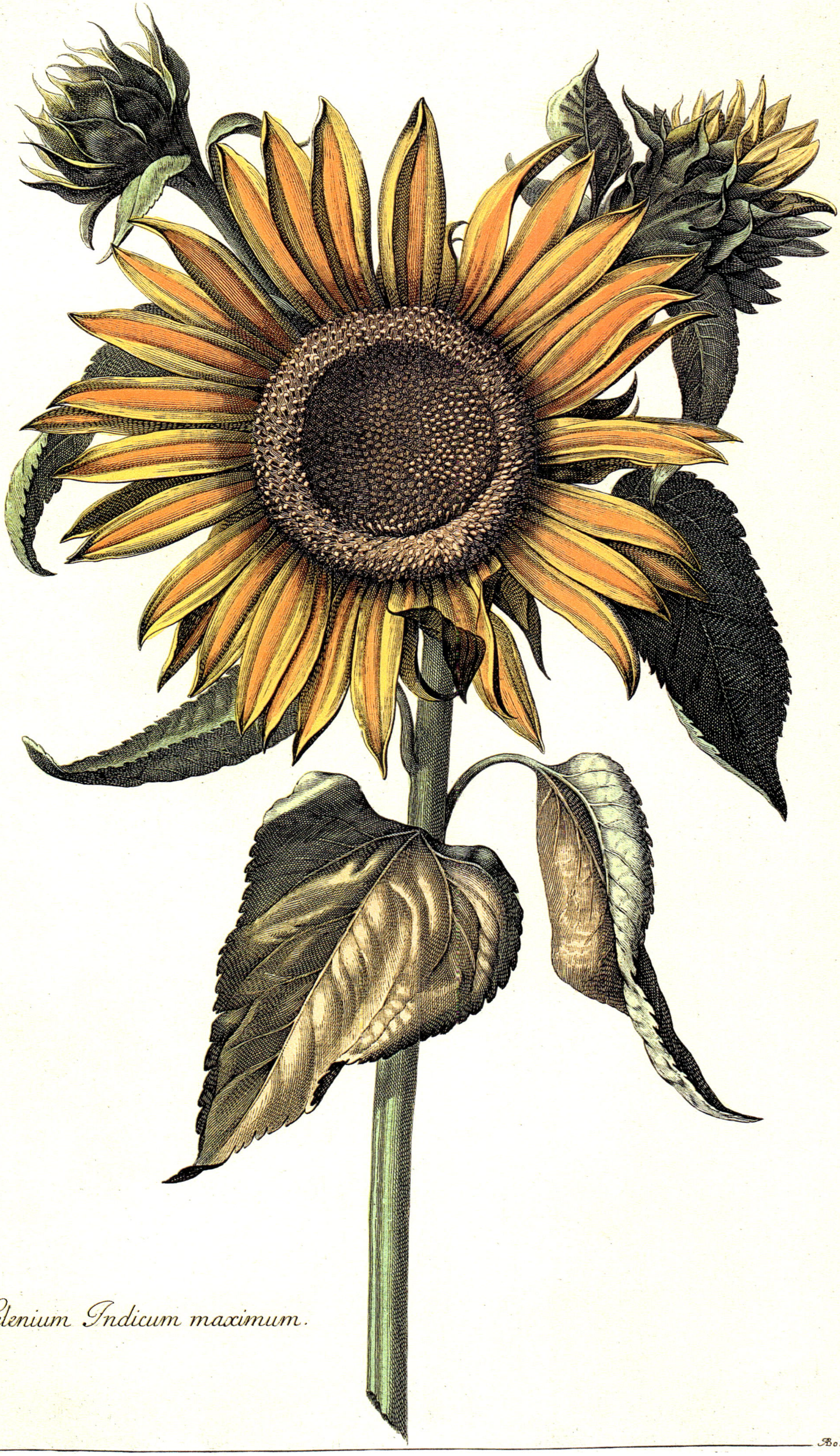

Helenium Indicum maximum.

A Bosse sculp.

Tanacetum parthenium (L.) Schultz Bip.

ASTERACEAE (EX-COMPOSITAE)

Feverfew, once known as a chamomile, is a typical example of the complexity of classifying certain plants. This engraving in fact shows a matricaria: *Matricaria parthenium* L. However, the plant was previously called *Pyrethrum parthenium* Sm. and *Leucanthemum parthenium* G.G. by Bonner (1934) and Costes (1937), then *Chrysanthemum parthenium* Bernh. by Fournier (1948). Finally, *Flora Europea* (1976) opted for *Tanacetum parthenium* Schultz, a tansy whose capitula have ligulate flowers, whereas the principal species, the common tansy (*Tanacetum vulgare* L.), does not!

Fortunately, throughout this long history, the plant kept its 'maiden name', making it possible to follow its trail. The Greek word *parthenos* means 'young girl' and also 'pupil' (of the eye), from the Latin *pupa* (young girl, pupil of the eye) which in turn gave rise to the French terms *poupée* (doll) and *pupille de la nation* (war orphan). The term 'matricaria', like *parthenium*, recalls the emmenagogue properties of the plant, which stimulates menstruation: from *matrix* (womb) and *mater* (mother).

It is of little consolation to know that the distinction between matricaria and chamomile is no easier, and even the great botanist Linnaeus (1707–78) confused them. For the sake of simplicity, chamomiles are *Anthemis*, mayweeds are *Matricaria*, while the feverfew shown is a *Tanacetum*.

The properties of these plants are similar but not identical. They all have a general digestive action (on the stomach, liver and intestine) and are emmenagogues, antispasmodics, febrifuges and diuretics. Indeed, it is impossible to tell which of these plants St Hildegard was referring to as *febrifuga* in the twelfth century.

Recently, feverfew has attracted renewed interest for its effectiveness in the treatment and prevention of headache and migraine. The wife of a senior health officer in the British mining industry was successfully treated with it on the advice of one of the miners and Dr Stewart Johnson then went on to carry out clinical trials and obtained very positive results.

Whatever these *camomilles* may be, they were once one of the most popular remedies used in France and many elderly people still drink a chamomile infusion in the evening to aid digestion.

Tanacetum parthenium (L.) Schultz Bip. Engraving : Nicolas Robert
National Museum of Natural History, Paris.

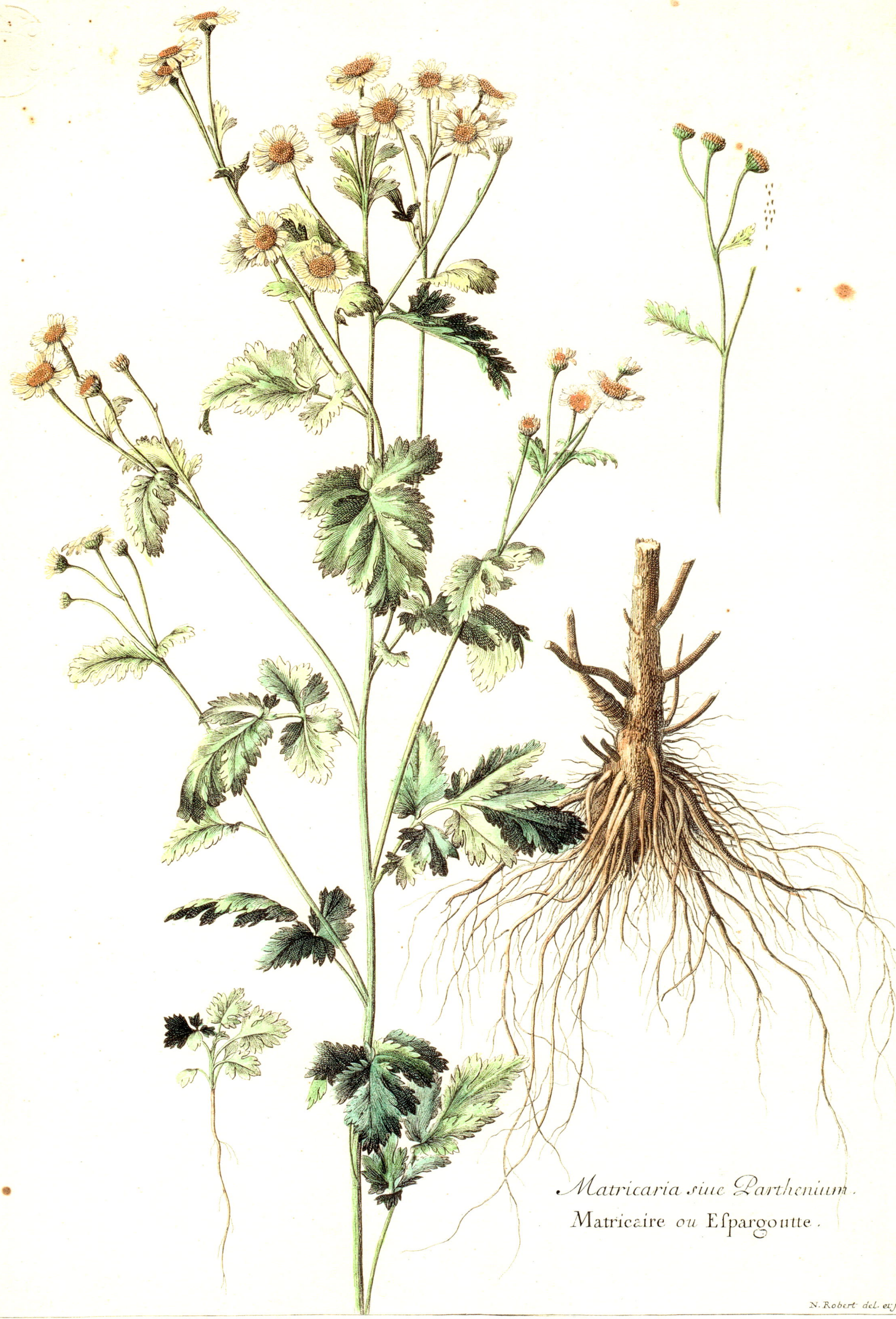

Matricaria siue Parthenium.

Matricaire ou Eſpargoutte.

N. Robert del. et ſculp.

Senecio doria L.

ASTERACEAE (EX-COMPOSITAE)

It is difficult to be certain exactly what plant is represented here. At first sight one might be tempted to think that it is *Senecio nemorensis*, the wood groundsel, but the caption – *Herba doria* – and the plant's resemblance to certain specimens of *Senecio doria* cast doubt on this. It would have been helpful to know the habitat from which the plant was taken – whether it was a damp situation or woodland – in order to identify it with more certainty.

Formerly known in Latin as *Senecio altissimus* Mill. and *Senecio carnosus* Lam., the golden ragwort is not very common in France, except in a few specific locations. It is found mainly in damp areas, meadows or marshland in the Mediterranean region, from the Alpes-Maritimes to the Aude, in the lower Rhône valley and the Camargue. It has also been found in some spots in the Ain region and was listed as a protected species in Languedoc-Roussillon by government order of 29 October 1997.

This handsome, hardy plant can grow to as much as 1.5 metres (5 ft) in height and forms quite large clumps, often growing in the company of the Montpellier thistle, the fig wort and horse mint. Its impressive leaves, almost entirely edged with small teeth, are fleshy and glabrous. The numerous, small, yellow capitula are densely grouped in composite corymbs at the top of the flower stalk. In the past, *Senecio doria* was often confused with the goldenrod.

In French its vernacular names include *séneçon charnu* (fleshy groundsel), *séneçon de Dorie* (Dorian groundsel) and *herbe des personnes âgées* (old people's herb). The latter name probably comes from the small hairy tufts on the fruit, typical of the senecios: from the Latin *senex* (old man).

One of the earliest major herbariums, that of Caspar Bauhin (1658), contains 2,360 plant specimens, including numerous members of the genus *Senecio*: *doria*, but also *alpinus*, *aquaticus*, *cineraria*, *fuchsii*, *coronopifolius*, *incanus*, *jacobaea*, and more.

Senecio jacobaea (common ragwort), *Senecio cineraria* (silver miller), *Senecio saracenicus* (broad-leaved ragwort) and *Senecio doria* (golden ragwort) are all powerful emmenagogues. They are dangerous and certain Mexican species can be fatal.

It should be noted that senecios, even the common garden groundsel, are not edible, despite their fleshy and tempting appearance. They contain pyrrolizidinic alkaloids which are highly toxic to the liver and can cause cirrhotic lesions.

Senecio doria L. Engraving : Nicolas Robert
National Museum of Natural History, Paris.

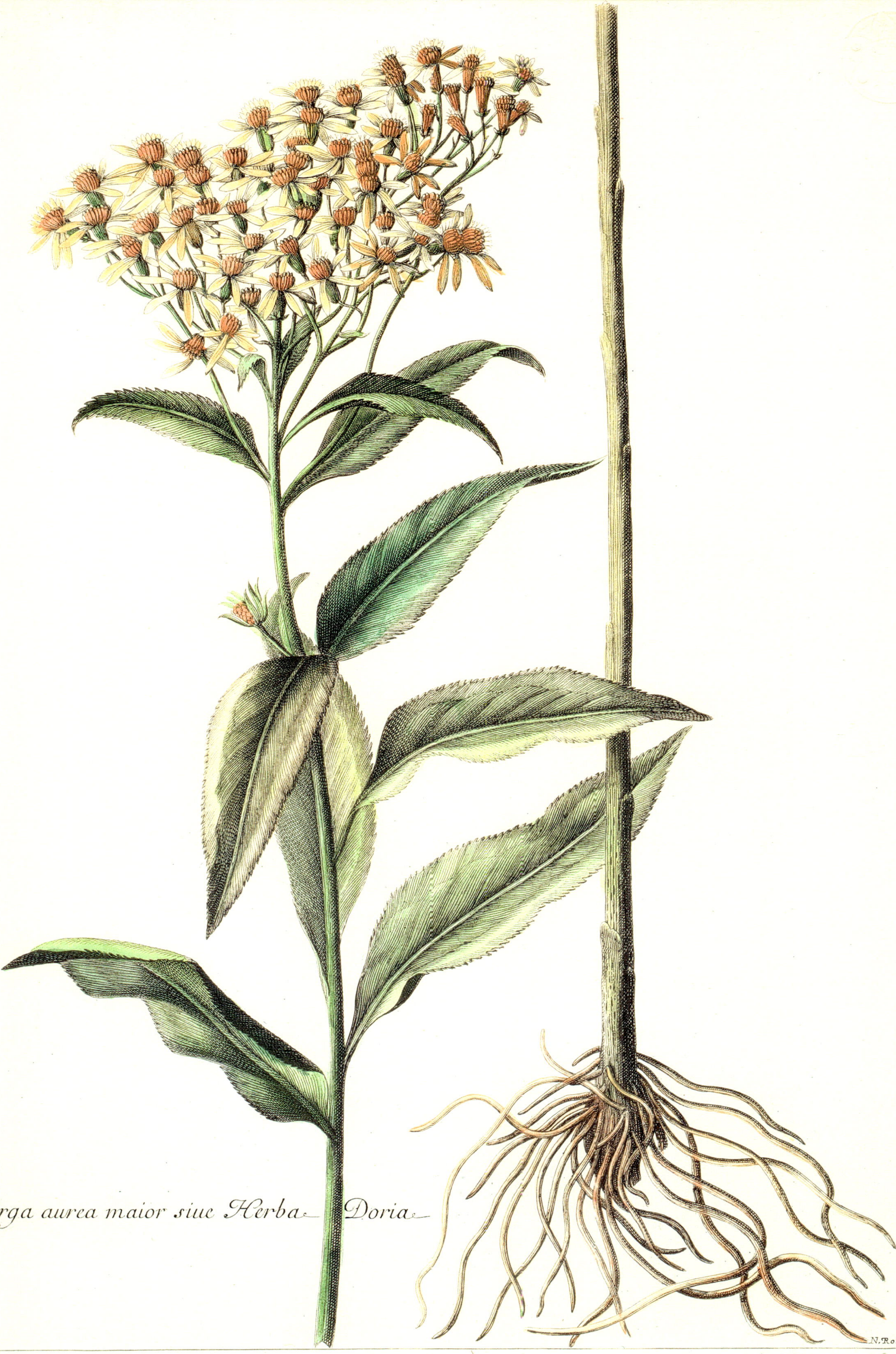

Virga aurea maior siue Herba Doria

N. Robert in. f.

Solidago sempervirens L.

ASTERACEAE (EX-COMPOSITAE)

The evergreen goldenrod is one of the major species of *Solidago* that originally came from northern America. It can grow to 2 metres (over 6 ft) in height in a favourable situation. The species does not appear in any of the classic floras. *Flora europea* (1976) only mentions it as being found as the subspecies *azorica* (Hochst) St-John in the Azores, where it has adapted by producing broader leaves at the base and smaller capitula. Kerguélen refers to it in *Index synonymique de la flore de France* (1993) but data concerning its location is not yet available. It seems possible that it may exist in the Charentes and Pyrénées-Atlantiques regions of France.

However, other North American species are so commonly found in the wild in France that they are listed as invasive in *Plantes invasives en France* (Muller, 2004). This is the case with *Solidago canadensis* Aiton and *Solidago gigantea* Aiton, introduced into England in 1648 and 1758 respectively. These two very similar species are often confused and colonise roughly the same habitats: slopes, embankments and waste ground. They are attractive plants and are therefore cultivated in gardens as ornamentals and as melliferous plants, which has assisted their dissemination.

The old name *Virga aurea limony folio* and the French terms *verge d'or* or *bâton d'or* were among the names commonly used to describe large plants with long, fine, yellow flowers. Known as the *verge d'or toujours verte* (evergreen goldenrod) in Quebec, it is one of the rare species of this genus to colonise the salty areas along tidal rivers and seashores. The level of salinity in the water affects the succulent nature (in the sense of 'water-filled') of its leaves, which are entire. The *Limony folio* appears to be related to the *Limonium* described by Matthiole (1500–77) – *Limonium* sp. – better known as the sea lavender, the basal leaves being very similar. This plant lives in very damp situations. *Limon* comes from the Latin *limus* (mud, silt) with the influence of the Greek *leimon* (damp meadow) and *limne* (marsh). The latter Greek word is also the origin of the French name *limnée* (lymnaea) a small, pulmonate, gastropod freshwater snail.

Other 'indigenous' *Solidago* grow in France, notably the species *virgaurea* L. This plant appears to be unknown to the Ancients, though they refer to other related species: Theophrastus (ca. 372–287 BC) mentions the *Chironion*; Pliny the Elder (23–79 AD) mentions the *leucophragis*; and Luigi Anguillara (1512–70) refers to *herba Scripta*. Arnaud de Villeneuve, known simply as Arnaud, (1238–1311 or 1313), was one of the first to mention its diuretic properties. Daléchamps (1615) has this to say of it: 'It is highly astringent and is an even stronger dessicative. Surgeons prescribe it to be taken as a beverage with fortunate success for those with bodily wounds and for fistulas, as it is exceptional at healing wounds and ulcers, both interior and exterior, and for stemming the flow of blood. Arnaud says that it is remarkable for urination and for breaking up stones. Taken as a beverage it controls stomach flow and also when used as a clyster.'

More recently it has been included in the Swiss vulnerary herb mix (which varies in number) known as *faltranck*, a German word originally meaning 'drink against falls'. It is recognised in contemporary herbal medicine for its popular use as a diuretic, vulnerary, astringent and antidiarrhetic due to the presence of terpenes and polyphenols. The plant is used in this way wherever it grows, even as far away as China where it is known in medicinal plant books as *Lieou kin nou ts'ao*.

Solidago sempervirens L. Engraving: Abraham Bosse
National Museum of Natural History, Paris.

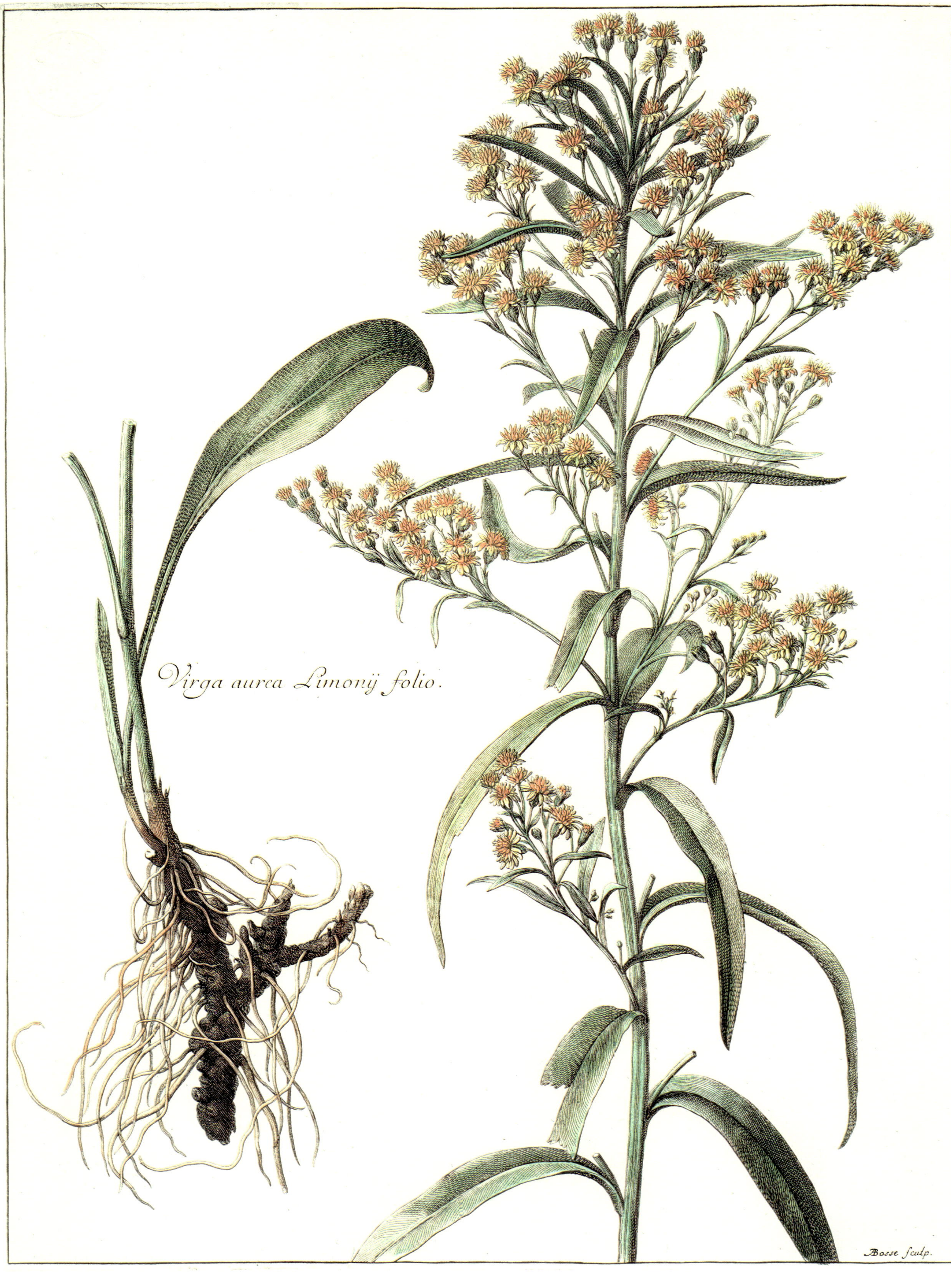
Virga aurea Limonij folio.
Bosse sculp.

Scolymus maculatus L.

ASTERACEAE (EX-COMPOSITAE)

The genus *Scolymus*, which resembles a thistle, is unusual because of the scale-shaped bracts of the receptacle surrounding the fruit. There are three species growing in France. Firstly, the spotted goldenthistle (*Scolymus maculatus*), which can grow to between 80 cm and 1 metre (32 and 39 in) in height and is found on uncultivated sites in the Mediterranean region. Its yellow-flowered capitula bloom in summer and it has deeply divided, white-veined leaves. Secondly, the large-flowered golden thistle (*Scolymus grandiflora* Desf.), which is very rare but can be found around Collioure and Banyuls in the Pyrénées-Orientales. Its large capitula of yellow flowers open from the end of May to early July.

The third, most common and best known, is the common goldenthistle (*Scolymus hispanicus* L.) which extends over a far wider area, from Brittany to the south of France, and throughout the countries bordering the Mediterranean, even as far as Iran. It is easily recognisable with its capitula of yellow flowers arranged in ramose clusters in the leaf axils and white-veined leaves. Its main root is long, thick and tapering.

It is commonly known in France as *chardon d'Espagne* (Spanish thistle), *épine jaune* (yellow thorn), *cardousse*, *cardouille*, etc. Wherever it grows, the roots and young leaves are eaten. In Tunisia, where it is known locally as *zirniz* or *cardousse*, the leaves and stems are eaten, along with the leaf nerves which are said to be excellent in couscous. Its use as a foodstuff dates back to Antiquity, as Daléchamps describes in 1615: 'As for Theophrastus's *Scolymus*, it has the advantage of a root that is good to eat, not only when cooked but also raw, and it is excellent to eat when the plant is in flower.'

The wild plant has been picked for centuries but has only been cultivated and improved quite recently. Poîteau and Vilmorin's *Le Bon Jardinier pour l'année 1848* offers some detail: 'A hardy or triennial plant ... Its long, fleshy root is used as a vegetable in Provence and Languedoc where the plant grows abundantly; in Montpellier it is known as *cardouille*. It is not however cultivated. It is picked wild from the fields, and as the central axis is usually woody, the root is split lengthwise and peeled away to leave the cortical part which is tied in small bunches and sold in the markets. Cultivation has improved the root quite noticeably.'

In Joigneaux's work of 1880, advice is found on the cultivation and preparation of this plant: 'It is grown for its roots, which are long and succulent. They are used in the following way: they are scraped clean and placed in vinegar diluted with water, they are then removed and put in boiling water with a little salt. Once cooked they are drained in a colander and prepared according to individual taste. They can be eaten with a poulette or a white sauce, like carrots, or with a sauce made from their juice, fat, flour, stock and seasoning. They can also be fried or eaten with salad ... The roots are ready to be lifted at the end of October but will overwinter very well in the ground in a climate such as that of Paris.'

Scolymus maculatus L. Engraving : Abraham Bosse
National Museum of Natural History, Paris.

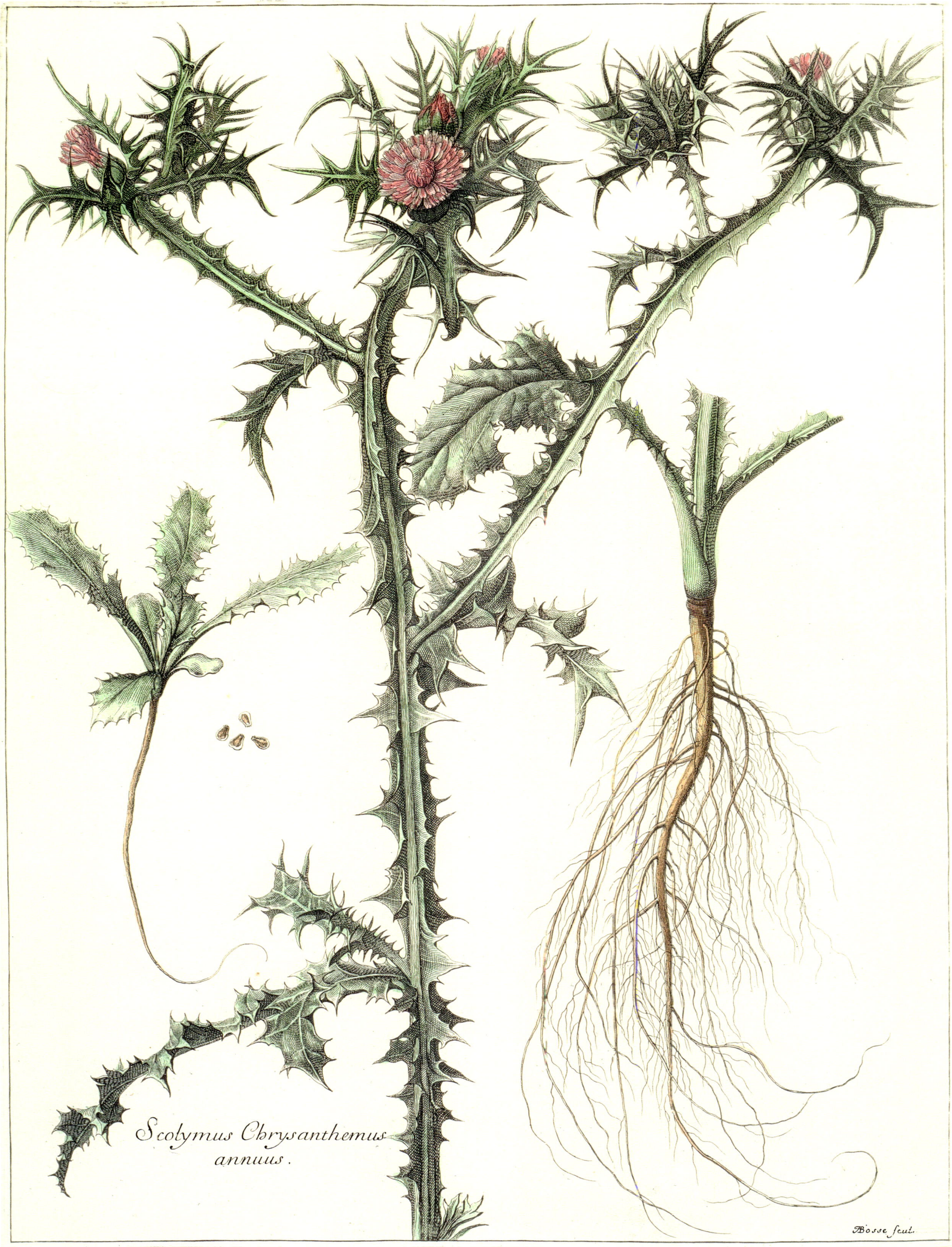
Scolymus Chrysanthemus
annuus.
ABosse scul.

Campsis radicans (L.) Seem.

BIGNONIACEAE

The bignonia, known to us since the seventeenth century, was cultivated in the Barberini Gardens in Rome. In 1638, Giovanni Battista Ferrari tells us that it was known as 'Indian jasmine' – a reference of course to the native indians of America.

The plant is called bignonia but its botanical name is *Campsis*. In the past, the genus *Bignonia* was an important group among the *Bignoniaceae*. Nowadays it is reduced to a single species, *capreolata*, along with other genera such as *Sariteae*, *Pandoreae* and *Clytostoma* – in fact this is true of 120 genera and 725 species, the majority originating in the tropics. The *Bignonia* which became *Campsis*, now contains two species: *radicans*, originating in North America, and *grandiflora*, which comes from China and Japan.

It was the Portuguese botanist Joao de Loureiro (1717–91) – an expert in Chinese flora and discoverer of the 'Chinese cabbage' (Brassica pekinensis or *haku-sai*) and of 'Pekin mustard' (*Sinapis pekinensis*) – who christened this plant *Campsis*, from the Greek *kampsis* (curved), a reference to its incurved stamens.

Bignonia is the name it was given by Tournefort in 1694, in honour of his patron, Jean-Paul Bignon (1662–1743). Bignon was first a priest at Saint-Quentin, then librarian to the king before, at the age of twenty-nine, becoming an honorary member of the French Academy of Sciences and, two years later, a member of the French Academy. In 1712 he published a novel entitled *Les Aventures d'Abdalla, fils d'Hanif*, which relates the adventures of a pious Muslim sent by his master, the sultan of the Indies, in search of a miraculous spring, the water from which would grant eternal youth.

The vernacular names for this plant come from its highly characteristic flower. It is known in French as *jasmin de Virginie* (Virginia jasmine) or *jasmin trompette* (trumpet jasmine) and in North America as the trumpet creeper or trumpet vine.

Its superb and abundant clusters of flowers blooming from July to September meant that it was soon adopted by gardeners as an ornamental to cover walls, arbours and pergolas. It is a vigorous creeper that can withstand temperatures in the order of -12° to -15° C. It can grow to 12 metres (40 ft) long with an annual growth of 1.5 metres (5 ft). A self-clinging creeper, it will climb on any kind of support, including trees and electricity poles. It also makes an excellent ground-cover plant.

However, the other side of the coin is that the vigour and hardiness of the trumpet vine have made it invasive in some parts of North America. It has colonised fields and roadsides and is considered a weed in the south-east United States. Yet it also tops the list of plants attractive to hummingbirds.

Campsis radicans (L.) Seem. Engraving: Nicolas Robert
National Museum of Natural History, Paris.

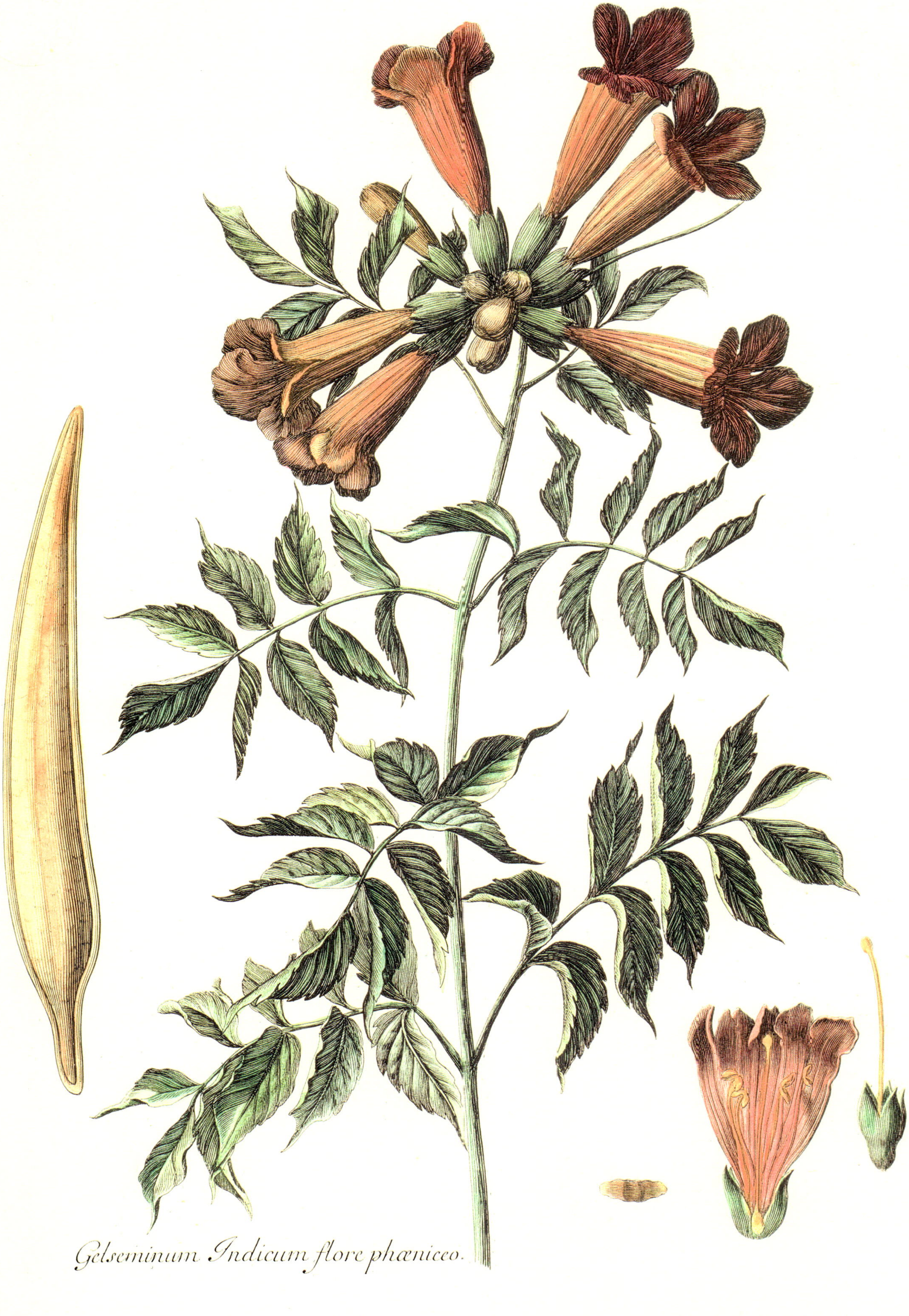

Gelseminum Indicum flore phœniceo.

N. Robert in. f.

Legousia speculum-veneris (L.) Chaix

CAMPANULACEAE

Although not a campanula, 'Venus's looking glass' is part of the campanula family. Its former French name – *doucette* – is curious. The plant was indeed eaten, like the more familiar *doucette* or *mâche* (corn salad or lamb's lettuce) – *Valerianella locusta* (L.) Laterrade var. *locusta* (ex-*V. olitoria*) – but though the leaves, which are picked when at the rosette stage, are pleasant to eat, it bears very little resemblance to this species.

The beauty of the flowers of this herbaceous annual, which are arranged in spreading, leafy corymbs, has earned it universal recognition as being worthy of Venus or womankind. Indeed, in French it is known as *Miroir de Vénus*; the Italians call it *Specchio-di-Venere*; the Flemish call it *Vrouwenspiegel*; and the Germans call it *Frauenspiegel* or *Venusspiegel*.

In the floras produced by Bonnier (1934), Costes (1937) and Fournier (1946), *Legousia* are known as *Specularia*. The species were named by Alphonse de Candolle (1806–93), son of Augustin Pyrame de Candolle. Claude-Auguste Durande, a renowned doctor and chemist, proposed to the Dijon Academy that the genus should be called *Legousia* in honour of Bénigne Legouz de Gerland, who founded Dijon's botanical garden in 1773. Jean-Jacques Rousseau had rejected a teaching post offered to him by Gerland and it was Durande (also author of *Flore de Bourgogne*) who accepted it.

As for the term *speculum* ... the first to see the beautiful, violet corolla with its white centre as a mirror was John Gerard, also known as Gerarde (1545–1612), an English physician and botanist who created the gardens for Lord Burleigh in London. His celebrated work, *The Herball or general Historie of plants*, was an adaptation of Rembert Dodoens' *Histoire des Plantes* (1554), published in 1597, containing more than 1,800 wood engravings, including one of the very first illustrations of the potato.

Many of the plants found in the fields at harvest time, like the *Legousia*, have virtually disappeared from floras, including the corn-cockle, the cornflower, the poppy and many others that were abundant at the time of Daléchamps (1615), who tells us that the corn-cockle (*Agrostemma githago* L.): 'Is Theophrastus's *Anthemon fueillu*. Theodore of Gaza calls it *Nigella foliosa*. Some know it as *Gitthago* and *Nigellastrum* ... The French call it *Nielle vulgaire*.' Of the cornflower (*Centaurea cyanus* L.) he remarks: 'Pliny says that one can recognise the flower of the *Cyanus* from its name, and *Blaptisecula*, because, as Ruel says, it causes problems for reapers and harvesters by getting caught in the cutting tool, known in Latin as *secula* (sickle). In French it is called *Aubifoin*, *Blauet* or *Bluet*.' Finally, Daléchamps has this to say about the corn poppy (*Papaver rhoeas* L.): 'In Greek the corn poppy or wild poppy is called ... that is to say *Pavot coulant* because of its fast-dropping flower. Apothecaries call it *Papaver rubrum*. The French *confanons*, *Ponceau*, *Coquelicoc* ... Modern writers claim that it cures inconstancy, stops women's monthly bleeding, stems nosebleeds if applied to an ulcer, and that a decoction made from it will reduce swelling of the genitals if they are soaked in it.'

Legousia speculum-veneris (L.) Chaix. Engraving : Louis de Chastillon
RMN, Louvre Chalcography Dept, Paris (modern print).

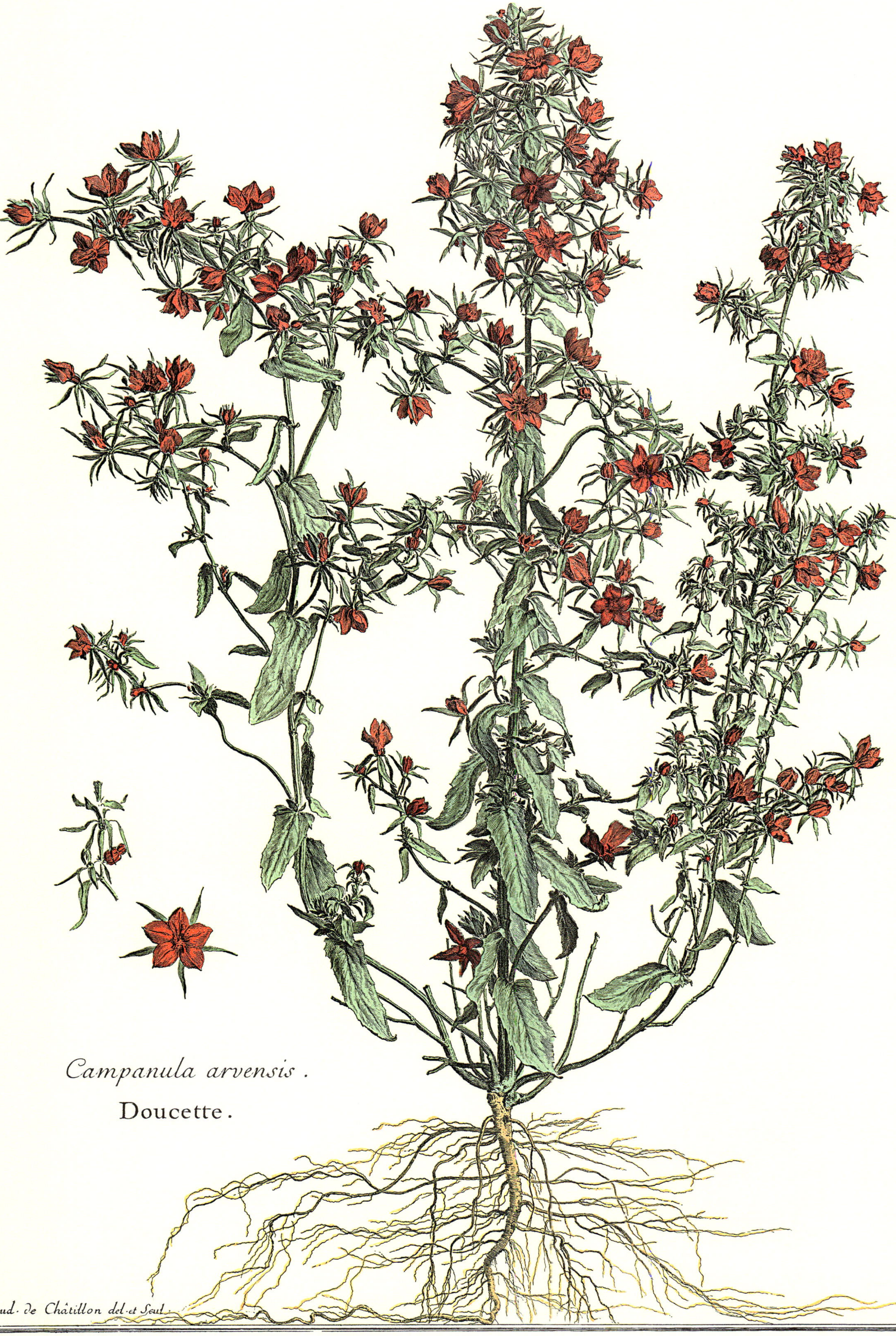

Campanula arvensis.

Doucette.

Lud. de Châtillon del. et Sculp.

Campanula thyrsoides L.

CAMPANULACEAE

The campanula shown here appears to be the yellow bellflower were it not for its blue colouring which is obviously due to a certain licence taken by its engraver, Nicolas Robert – old engravings often include examples of 'abnormally' coloured species. If we consult the floras compiled by Bauhin (1658) and Tournefort (1700), it is clear that it is in fact *C. thyrsoides* L.

Botanically speaking, a thyrse is a pyramid-shaped inflorescence made up of composite clusters of flowers. Well-known examples are the thyrses carried by bacchants during the wild and mystic festivals known as Bacchanalia: 'The bacchants, dishevelled, holding black masks, sway to the sound of the music, bunches of grapes hanging at their brows ... beating themselves with thyrses' (Gustave Flaubert, *The Temptation of St Anthony*, 1856). What is being described here is a symbolic attribute of Dionysus (Bacchus, god of the vine and of wine) – a stick entwined with stems of vines and ivy and topped by a pine cone or vine leaves.

Campanulas are characterised by their bell-like flowers. The term *campanula* has been used in botany since the seventh century and is borrowed from the low Latin word *campana*, which means both a steelyard balance and a bell. For centuries the word *campane* was used for *cloche* (bell) until the latter term, introduced by Irish evangelist monks, came into definitive use in the seventeenth century.

One legend tells how St Paulin, at the time bishop of Nola (in Campagna), was in the fields sleeping off the wine he had drunk when he believed he heard celestial music. All around him campanulas were swaying in the breeze and this gave him the idea of making bells in the shape of these flowers. Hence, originally, the names *campana* and *nola*. Historically, five different types of bell were used in convents: the *squilla* – used in the refectory; the *cymbalum* – used in the cloister; the *nola* – in the choir; the *campana* – in the bell tower or steeple; and the *signum* in the other towers.

Campanulas have mainly colonised the northern hemisphere, adapting to many different kinds of biotopes: dry areas of scrubland and plain; areas of scree and fallen rock; high mountain pastures; rocks and cliffs, etc.

Few have been used medicinally, but some are edible. The rampion bellflower has long been cultivated and produces a delicious salad leaf. It is still picked in the south of France and in warm, dry spots at medium altitude in the mountains. In the past it was eaten throughout France, from north to south.

This particular campanula, called *Rapunzel-Glockenblume* in German, was so well known that it gave its name to a fairy tale recorded by the brothers Jacob and Wilhelm Grimm and published in 1857. It is the story of a pregnant woman who dreams of eating 'fine rampions, with leaves so green, and so glossy, so fresh and so appetising that her mouth begins to water'. The problem is that this irresistible salad leaf is growing in the garden of a witch ... and her child is born under a curse. The child's name is Rapunzel.

The yellow bellflower was cooked and eaten in certain areas of the Alps, a practice which, after endangering the species, has now virtually disappeared. Fortunately this means that it can still be enjoyed and admired in mountain areas in all its glory. As Daléchamps (1615) says, 'this plant is rare, but that makes it all the better to see'.

Campanula thyrsoides L. Engraving: Nicolas Robert
RMN, Louvre Chalcography Dept, Paris (modern print).

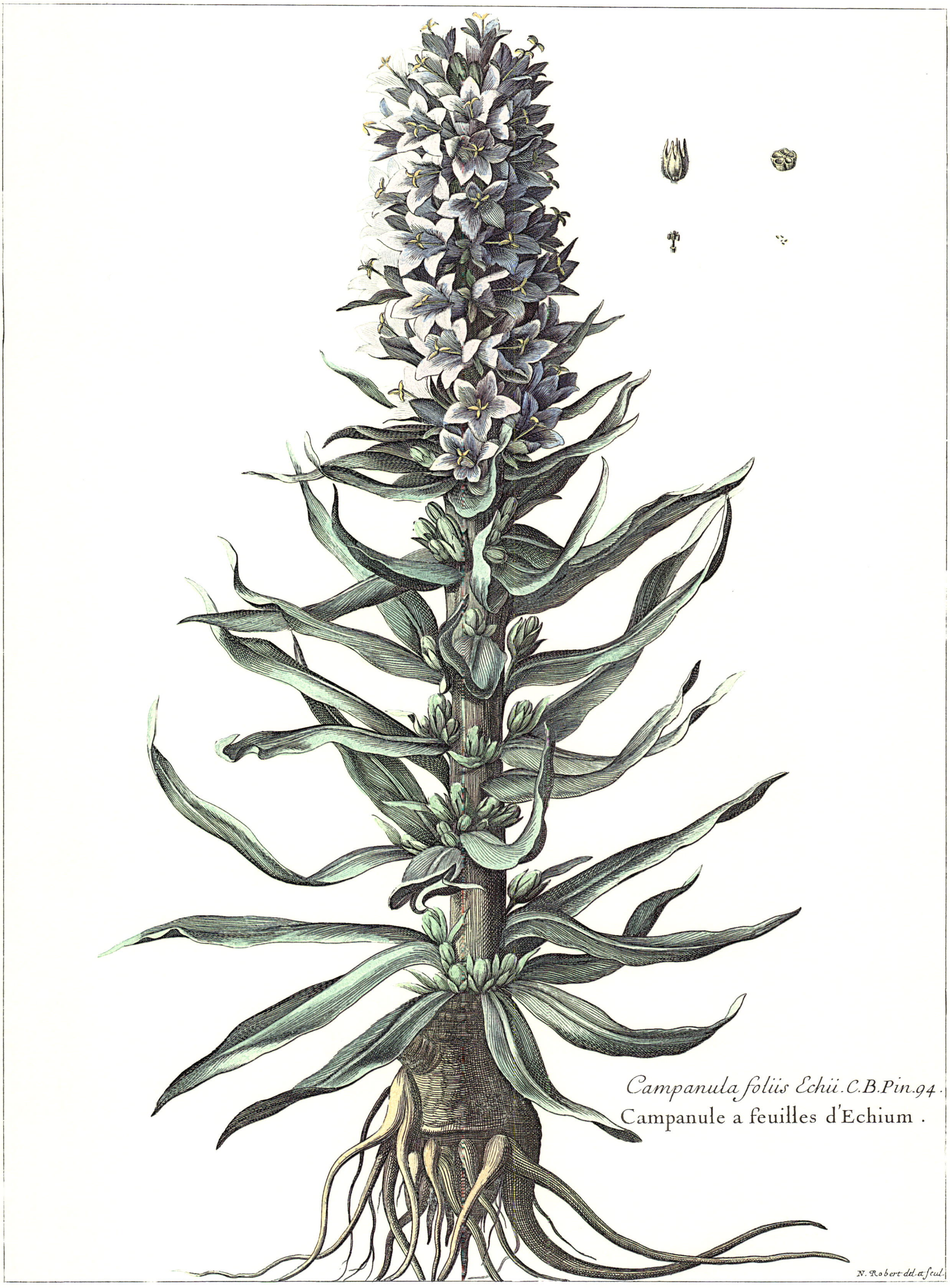

Campanula foliis Echii. C.B. Pin. 94.

Campanule a feuilles d'Echium.

N. Robert del. et sculp.

Arisaema triphyllum (L.) Schott

ARACEAE

Arisaema triphyllum, also known as *Arisaema atrorubens* (Ait.) Blume and *Arum triphyllum* L., is an herbaceous plant originating in southern Canada and the eastern side of North America. In its preferred habitat of cool, damp woodland and river banks it can be found growing abundantly in some areas. Denis Dodart (1676) tells us that 'according to Caspar Bauhin it was brought from Brazil in 1614. It has been brought back to us recently from Canada.'

The inflorescence consists of a greenish purple spathe with brown or white stripes and a flattened limb which suddenly broadens above the tube. This encloses a clavate spadix, the reproductive organ, which is dark purple in colour.

The plant is generally dioecious, which means that there are specimens with an entirely staminate spadix (male) and others with an entirely pistilate spadix (female). Nevertheless, specimens are also found with a mixed spadix, like European arums, which have staminate flowers in the upper part and pistilate flowers in the lower part. Proportionally, there are around three male plants to every two female plants and one mixed plant in every ten. In fact, this *Arisaema* only flowers after four years and the first inflorescence contains only male flowers. After a number of years an exclusively female inflorescence appears and then will remain so during the following years.

The usual nectar-gathering insects take no interest in this plant, except for gnats – which have the evocative name *brûlot* (fiery) in French. These formidable little Canadian flies are bloodsuckers and though, like mosquitoes, they go unnoticed as they begin to feed, they leave a burning sensation on the skin. Little is known about the pollination of *Arisaema* but some studies are underway on this subject and should prove enlightening. If the life of the insects that pollinate these *Arisaema* is anything like the incredible lives of those that pollinate the European arum – moving from one plant prison to the next – it should certainly prove interesting.

The handsome scarlet-red berries of the plant are poisonous to man, like those of the arum, but are eaten by birds, squirrels and woodmice. In Europe, and in France in particular, cases of children being poisoned by these attractive red fruits are recorded every year.

There is no immediate burning sensation when juice from the rhizome touches the skin. It begins to be felt after a few minutes and is caused by the presence of raphides – tiny needle-shaped calcium oxalate crystals. Once filtered, the juice is no longer an irritant. In fact, these rhizomes, which are rich in starch, are edible once dried and boiled. They were once used in popular medicine to treat 'weakness of the blood and stomach problems'. The indians cured colic by drinking water in which *Arisaema* rhizomes had been steeped. In France, *Arum triphyllum* appears only in the homeopathic register, where it is listed for the treatment of ear, nose and throat problems.

The genus name *Arisaema* comes from the Latin *arum*, which in turn comes from the Greek words *aris* and *arisaron* (hood) and *haima* (blood). In Quebec it is commonly known as the 'little preacher' as the upper, folded back part of the spathe resembles a pulpit and the spadix a priest or clergyman delivering a sermon. The English name – Jack in the pulpit – is even clearer.

The Iroquoi indians called the plant *Kah-a-hoo-sa*, which means 'cradle' or 'crib', as the shape of the spath resembles that of the *nagane*, a kind of pouch used by indian women to carry their baby on their back.

Arisaema triphyllum (L.) Schott. Engraving : Nicolas Robert
National Museum of Natural History, Paris.

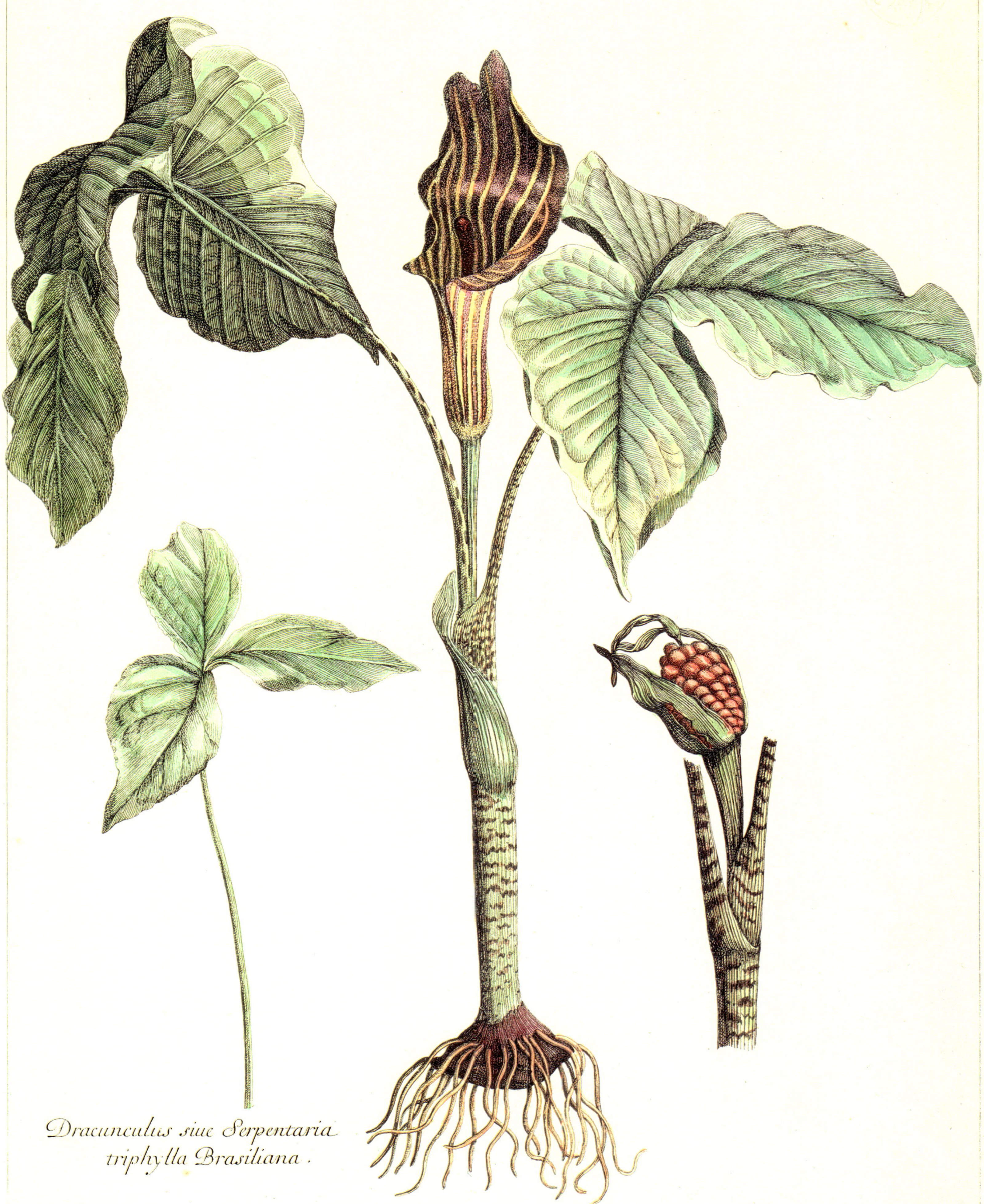

Dracunculus siue Serpentaria triphylla Brasiliana.

N. Robert in. f.

Lilium candidum L.

LILIACEAE

As a baby, very little was required of Herakles to gain immortality: he simply had to drink the milk of his sworn enemy Hera, wife of Zeus. Hermes was asked to place the child gently on the breast of the sleeping goddess, but Herakles suckled so strongly that Hera awoke and pushed away the infant she so despised. He however clung to her breast with such force that a large quantity of milk spurted forth and across the starry sky. This was the origin of the Milky Way. A drop of her milk fell to earth and produced an immaculate white flower – that at least is one of the many legends surrounding the lily.

For centuries this majestic plant has been considered a symbol of purity and virtue, hence the vernacular names associating it with the mother of Jesus, such as the Virgin lily, the Madonna lily, as well as the St Anthony's lily and in French *lis candide* (guileless lily) and *lis* or *rose de Junon* (Juno's lily/rose). The bird sacred to Juno (Hera in Greek) was the peacock, which represented the eyes of Argos, 'the watcher'; and her sacred plants were the helichrysum, the pomegranate and the lily.

The white lily appears to be the oldest ornamental plant in cultivation. It figures on a Cretan vase dating from the seventeenth century BC, found at Knossos, and on ancient Egyptian papyruses and frescoes. The lily was venerated by many ancient peoples from the eastern Mediterranean, including the Assyrians, Phoenicians and Hebrews. Among the latter it was known as *shôshannah* or *schuschan*, which became the forename Suzanne or Susan, and among the Latin peoples it was *lilium*, from the ancient Coptic *krêri* and *klêli*. Given this history, it is believed that the plant originated in this region, where is still grows wild today.

One of the remedies most commonly used in France in the twentieth century, and probably earlier too, was based on lily petals steeped in alcohol. It was remarkably effective when applied to infected wounds and also when used to promote the healing of abscesses and splinters. The petals were also steeped in olive oil and used to treat burns.

The white lily is still widely cultivated as a garden plant. It can be increased from seed and from bulb scales – the scales are carefully removed and planted upright in prepared soil. At the base of the scale a small bulge appears, followed by some roots and then a small bulb. All lilies can be easily increased by this method, including the rare orange-coloured lilies with and without bulbils, the martagon lily, the Pyrenean lily and the pompon lily.

In France the air of noble majesty emanating from this plant made it popular with royalty. In 1211, the *fleur de lis* appeared for the first time as a motif on the seal of Prince Louis, later Louis VII. Charles V adopted the emblem of three lilies on a blue background in 1375 in honour of the Holy Trinity, thus cleverly associating royalty with Christianity. There was some dispute over the origin of this particular lily. In fact it was not a lily but a marsh iris (*Iris pseudacorus* L.), with a trilobed corolla – a flower much loved by Louis VII. It is thought in fact that the term *fleur de lys* is a corruption of *flor de Loys* (Louis' flower).

Lilium candidum L. Anonymous engraving
RMN, Louvre Chalcography Dept, Paris (modern print).

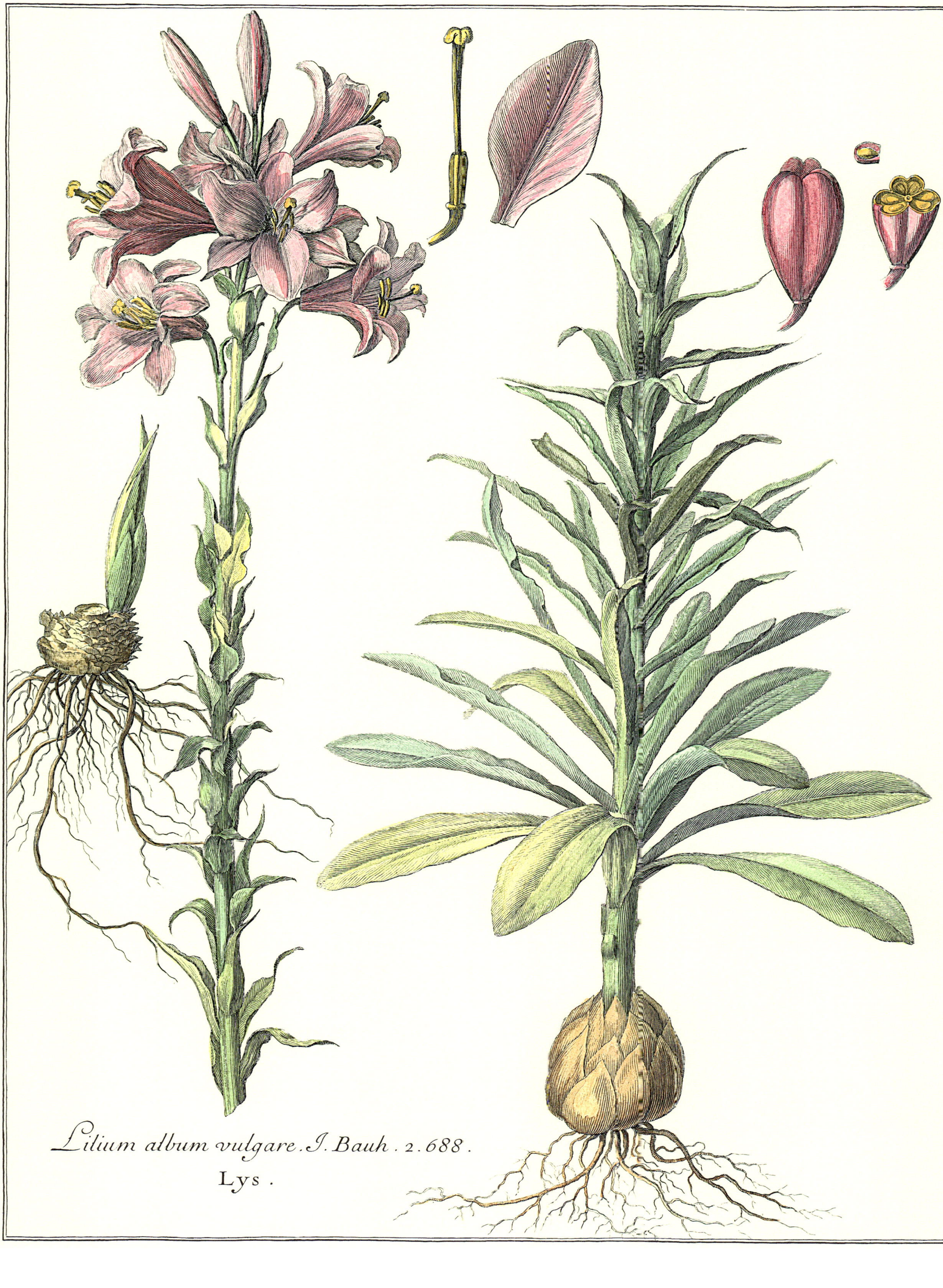

Lilium album vulgare. I. Bauh. 2. 688.

Lys.

Tulipa sp.

LILIACEAE

The late sixteenth century and the seventeenth century were a glorious time for florists, a time when tulips were bought and sold for incredible sums of money.

Augier Busbeck (1522–92) – who, as was the custom among many distinguished people at that period, latinised his name to Augerius Ghislenias Boubequius – was a Dutch diplomat and botanist. It was he who first discovered the tulip in Constantinople in 1555–56, while ambassador of the Austrian Hapsburgs at the court of Suleiman I, known as 'the Magnificent'. Busbeck sent some tulip bulbs to Charles de l'Écluse, known as Clusius (1526–1609), who was director of the imperial gardens in Vienna at the time. Clusius became professor of botany at the university of Leiden in Holland in 1580 and introduced tulip cultivation there. Holland and soon the whole of Europe became obsessed with this flower in a way unprecedented in the history of botany. It became a symbol of beauty, rarity and hence of wealth, rather as the orchid was before people understood how to cultivate it.

The average annual income of a labourer in Holland at that period was 150 florins, yet tulips were being traded for phenomenal amounts: in 1623 the very rare *Semper Augustus* changed hands for 1,000 florins; in 1625 for 2,000 florins; and in 1637 for 5,500 florins.

Tulips were fought over by the wealthy. 'Tulipmania' suddenly broke out, in much the same way as an economic bubble develops today. In fact, what people speculated on were the species, the rare varieties and new introductions. Given the high prices and the fact that a large bulb would produce a large bloom, bulbs were carefully measured, and weighed to the nearest grain. The grain was an old unit of weight equivalent to 0.0532 grams, i.e. a 72nd part of a gross or a 24th part of a scruple, which is itself a 24th of an ounce, which is a 16th of a pound, the value of which varied from one region to another. In the famous La Fontaine fable, the four 'grains' of hellebore that the tortoise was supposed to take as a purge were in fact 0.21 of a gram of this plant.

In 1636, Adriaen Roman published a pamphlet describing how a weaver gave up everything to become a florist. It enlightened the small world of speculators about the irrationality and fragility of the system. A tulip 'crash' soon followed and the bubble rapidly burst collapsing in on itself with a 99% fall in prices and leaving a string of ruined, desperate individuals.

Today there are at least 125 species of tulip in the world and around 5,600 varieties. Most garden tulips are produced from *Tulipa gesneriana* L. In France, almost all wild tulips are protected.

As for the name 'tulip', Daléchamps (1615) has something to say about it in his article on the *Tulipan*: 'Modern authors have called this flower, which is foreign, by a name which is also foreign, which is *Tulipan*, a Turkish name. The Slavs and the Turks use this to describe what they customarily wear on their head; and they have given it this name because these flowers, when turned upside down, resemble this type of headwear.' The words tulip and *tulipan*, come from the Turkish *tülbend-lâle* (*lâle* meaning flower in Turkish), which in turn is borrowed from the Persian *dul-i-band* (turban) because of its shape and the beauty of its colours.

Tulipa sp. Engraving : Louis de Chastillon
RMN, Louvre Chalcography Dept, Paris (modern print).

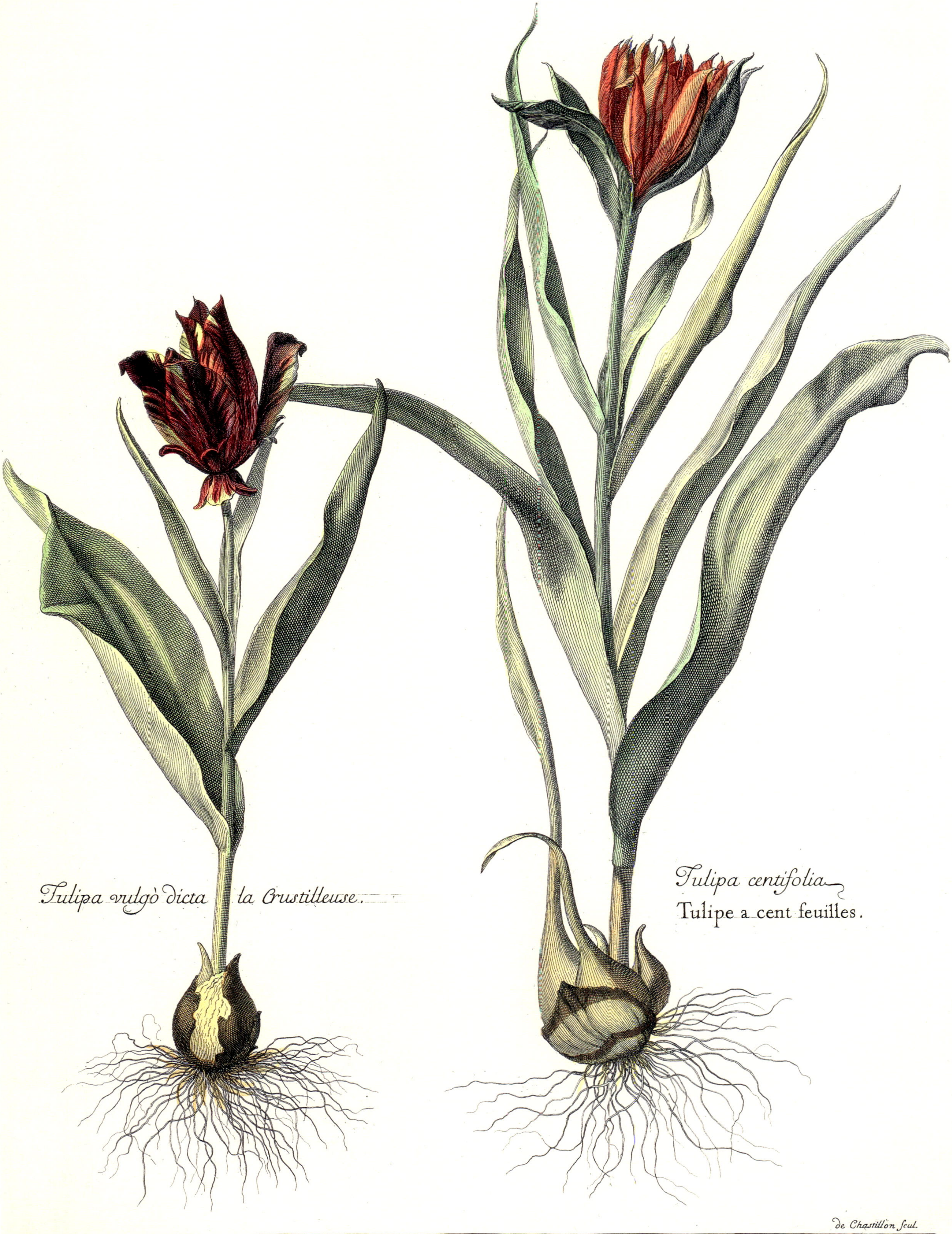
Tulipa vulgò dicta la Crustilleuse.
Tulipa centifolia
Tulipe a cent feuilles.
de Chastillon sculp.

Leucojum aestivum L.

AMARYLLIDACEAE

The *Leucojum*, previously known as *Leuconium*, is one of nature's treasures. Very rare, it is better known in France as the *nivéole* and in England as the summer snowflake. It is generally a small and unobtrusive plant with elegant, white, bell-shaped flowers. Its Latin name comes from the Greek words *leucos* (white) and *ion* (violet). *Nivéole* is a reference to its colour being white like snow, but also to the fact that some species flower at the end of winter. The root of this word is also found in *nivôse* – the month of snow (in the French republican calendar created by Fabre d'Eglantine in 1793) which began around 21 December.

In fact, the single-flowered spring snowflake and the snowdrop are very similar in appearance. From the visible part of the plant they can be distinguished by their petals, which are larger and even in the snowflake, uneven in the snowdrop. The summer snowflake shown here has a terminal cluster of three to seven small, white flowers, all emerging from the same longitudinal fissure of the bract enclosing the flower buds. In some areas of France it is known as the *nivéole à bouquet* (the cluster snowflake).

French flora contain seven species, including two subspecies for *Leucojum aestivum*: subsp. *aestivum* and subsp. *pulchellum*. The other species are *automnale*, *fabrei*, *longifolium*, *nicaeense*, *roseum* and *vernum*.

The snowdrop, *Galanthus nivalis* L., from the Greek *gala* (milk) and *anthos* (flower), though delightful is nevertheless toxic. The *Galanthus*, along with the *Amaryllis*, *Narcissus*, *Lycoris*, *Crinum* and *Leucojum*, belong to the family *Amaryllidaceae*, rich in toxic alkaloids.

Although to the layman the bulbs may look like small onions, they are highly dangerous and accidents still occur all too frequently. The bulbs of certain species contain more than a hundred alkaloids, the most common of which is lycorine, also known as narcissine and galanthidine, which is too toxic to have any use in medicine. Only galanthamine is used for its action on the sympathetic nervous system.

In their country of origin, the majority of these bulbs were once used to poison arrows. It is important to remember that however charming a plant may appear, it can nevertheless conceal a deadly poison. So, when picking flowers, always beware of judging a plant harmless because it looks innocuous.

Leucojum aestivum L. Engraving : Nicolas Robert
National Museum of Natural History, Paris.

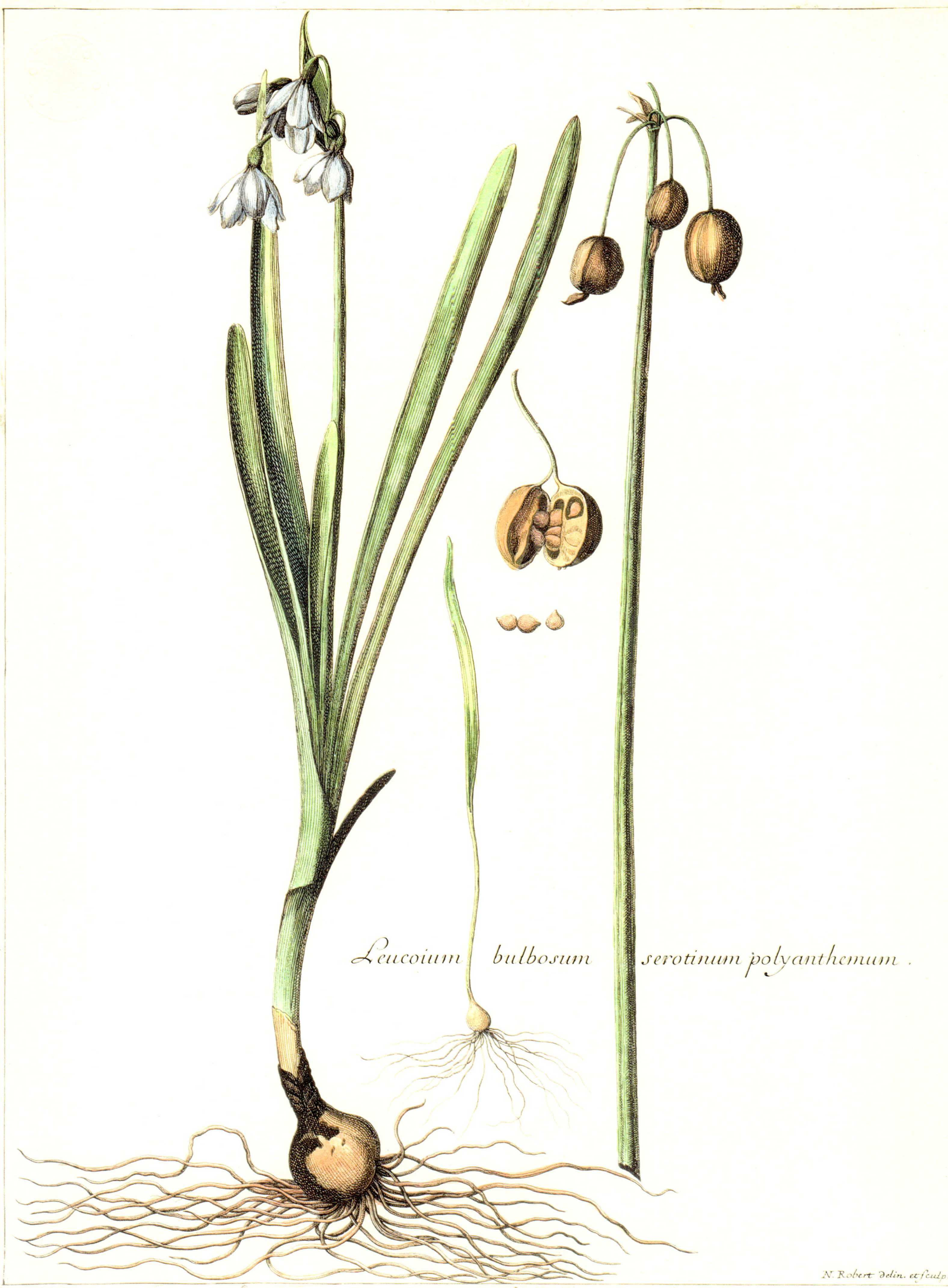

Leucoium bulbosum serotinum polyanthemum.

N. Robert delin. et sculp.

Narcissus pseudonarcissus L.

AMARYLLYDACEAE

For most people, anything that resembles a narcissus at any distance is a 'daffodil', regardless of whether it is the poet's narcissus, *pseudonarcissus narcissus* (the wild daffodil) or the *narcisse des garrigues* (the Mediterranean scrubland narcissus). However, a true daffodil in fact exists: *Narcissus jonquilla* L. The French term *jonquille* and the English *jonquil* refer to the form of the leaves, which are similar to those of the *jonc* (rushes).

The narcissus appears in the Greek myth of Tiresias, the famous seer of Antiquity who experienced life both as a man and a woman, and who carried a wooden staff of second sight made from male dogwood. He predicted to the parents of the newborn infant Narcissus, that if their son saw his reflection he would die. One can imagine the parents' distress at discovering that their son's survival depended on removing anything that might allow him to see his reflection.

Narcissus grew up unaware of how he looked though he was a young man of great beauty with whom both goddesses and mortal women fell in love. However, he remained unaffected by matters of the heart and moved among his admirers without even noticing them. One pretty nymph called Echo was so smitten with Narcissus that she stopped eating, drinking and sleeping. She became translucent, then transparent and eventually so weak that she barely had the energy to pronounce the final syllables of words. In her sadness she withdrew into the mountains where she is sometimes heard to echo our calls.

Eventually, all these unfortunate young women went to complain to the goddess Demeter, who decided to punish the proud young man. One day when he was out walking, the handsome Narcissus felt an unquenchable thirst at the very moment he came upon a spring – a magic one, naturally. When he leant over to drink, he discovered such a wonderful face looking back at him from the water that he could not remove his gaze. When he went to caress it the water became disturbed and when he cried each tear created a ripple. Days passed and Narcissus remained there until he eventually died. It is said that on the Styx (the river of the Underworld) he continued to seek the beloved features he had once admired. And at the spot where he died a flower emerged, its head bowed over in humility.

The word 'narcissus' is linked to the Greek *narké*, which also gave us the word 'narcotic'. The fragrance of some narcissi is so heady and overpowering that people who are sensitive to it can find it intolerable. The plant is also toxic. In fact cases of allergy have been reported simply from contact with the bulb and leaves.

Narcissus pseudonarcissus L. Engraving : Nicolas Robert
RMN, Louvre Chalcography Dept, Paris (modern print).

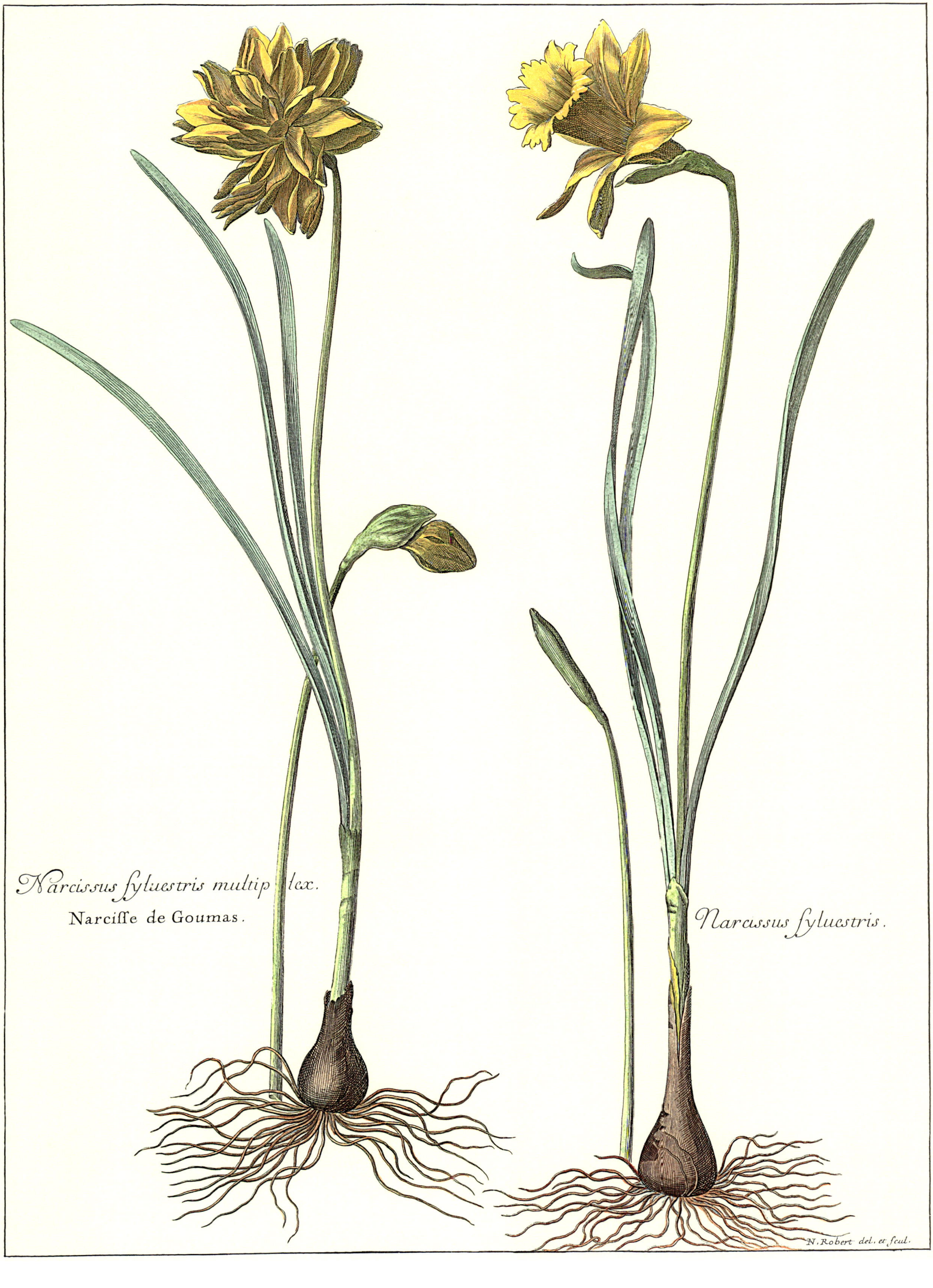
Narcissus sylvestris multiplex.
Narcisse de Goumas.
Narcissus sylvestris.
N. Robert del. et sculp.

Nerine sarniensis (L.) Herb.

AMARYLLYDACEAE

The nerine, a relative of the garden and pot-grown amaryllis, is finer and more elegant with its terminal clusters of long, slender flowers. Botanically speaking, the amaryllis – a bulbous plant originating in South America – in fact belongs to the genus *Hippeastrum*.

The genus name comes from the Nereids. These sea nymphs (fifty or a hundred in number, depending on the source) were the offspring of Nereus and Doris. Nereus, also known as 'the old man of the sea', was the son of Pontos (the great sea) and of Gaia (the earth). Doris was the daughter of Ocean, a personification of the sea enclosing the world and father of all rivers.

The beautiful and benevolent Nereids lived at the bottom of the sea in their father's palace, where each had a golden throne. They spent their time swimming with hippocamps and dolphins, their hair floating out across the waves; they also enjoyed spinning, weaving and singing.

They are not to be confused with the Sirens, who were sea demons and fewer in number, though there were at least three of them: Parthenope, Leucosia and Ligia. They attracted sailors with their singing, causing them to crash into the rocks at the entrance to the Strait of Sicily and then devoured them. In the familiar story of Orpheus, it was his singing, which surpassed that of the Sirens, that allowed the Argonauts to continue their journey unharmed. They also appear in the story of Ulysses who asks his companions to tie him to the mast of their ship so that he can hear the Sirens but resist their call: an experience that almost drove him mad.

There is often some confusion in our minds because the Sirens of Greek and Roman Antiquity had the head of a woman and the body of a bird, whereas those associated with Nordic myths had the tail of a fish. The story of the 'little mermaid', with which we are all familiar, is a combination of the two myths – a charming Nereid with a fish's tail.

So what is the link with the nerines that originated in South Africa? Some of these reached Europe by sea. The botanist Robert Morison (1620–83) recounts how, during his lifetime, a ship laden with bulbs of this plant was shipwrecked off the coast of Guernsey and that the plant took up residence there, hence the name Guernsey lily.

The botanist Jacques Philippe Cornut mentions the plant in 1635 with the name it is given on this engraving: *Lilionarcissus Japonicus rutilo flore*, which later became known as *Amaryllis sarniensis* L.

Nowadays, many species and varieties of nerine are available in gardening shops, all of them very elegant. Elegant ... but every bit as dangerous as Parthenope, Leucosia and Ligia as their leaves and bulbs contain cytotoxic alkaloids!

Nerine sarniensis (L.) Herb. Engraving: Nicolas Robert
RMN, Louvre Chalcography Dept, Paris (modern print).

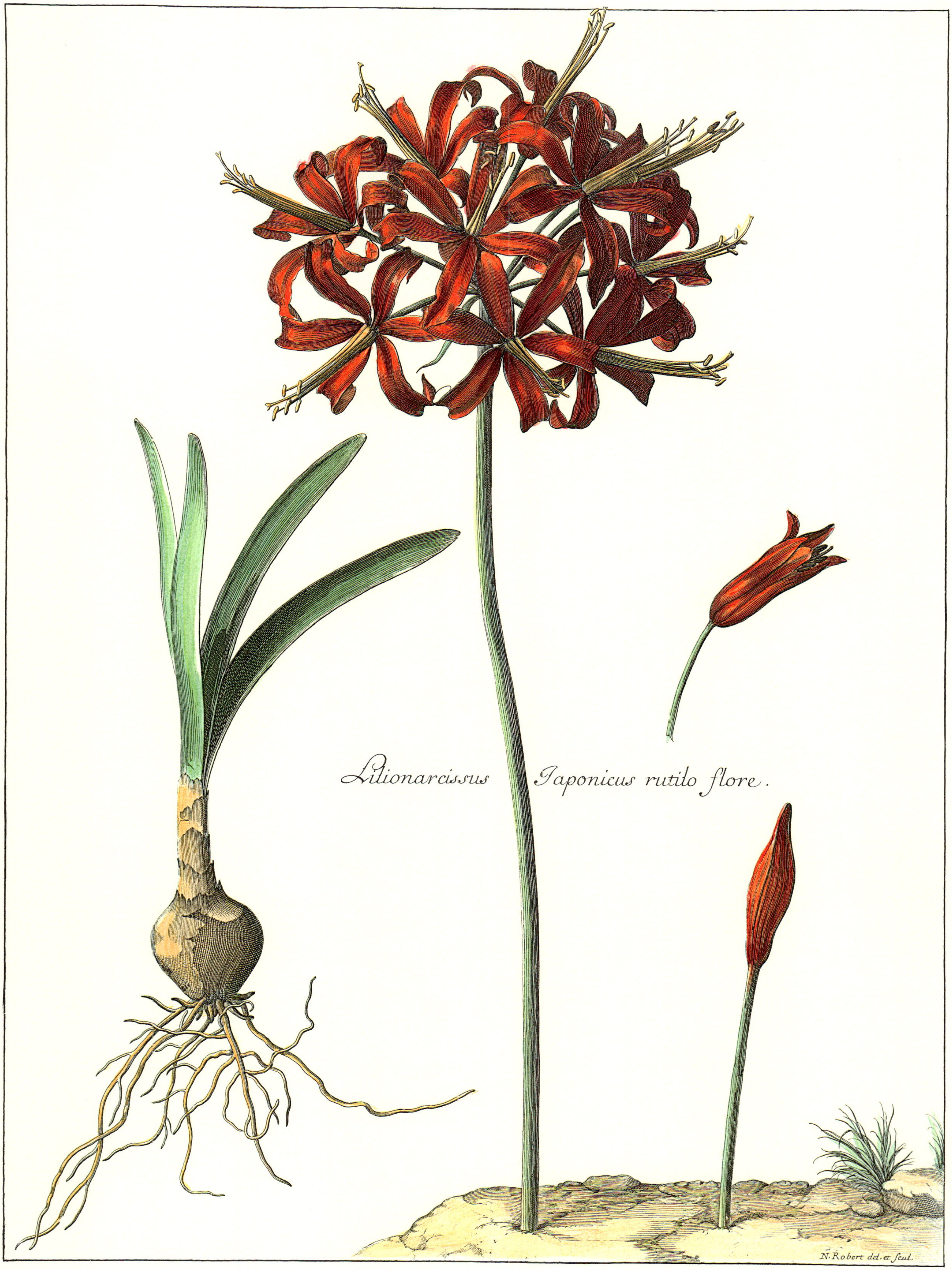
Lilionarcissus Japonicus rutilo flore.
N. Robert del. et Sculp.

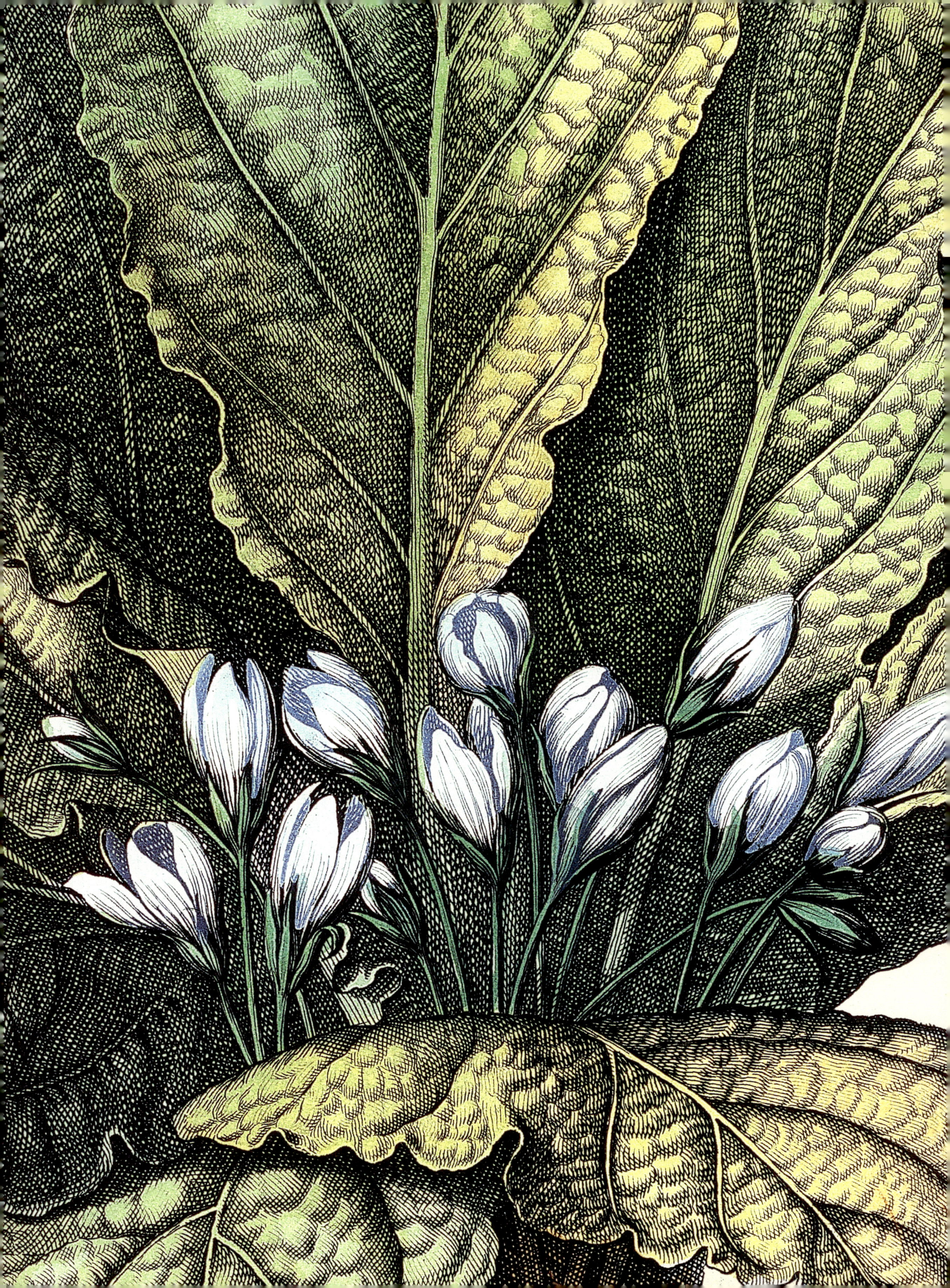

Bibliography

ALLORGE, Lucile and IKOR, Olivier, *La Fabuleuse Odyssée des plantes,* Paris, JC Lattès, 2003.

BAUHIN, Caspar, *Theatri Botanici sive Historiae Plantarum,* Apud Ioannem Kônig, 1658.

BELOT, J., BORNAREL, P. et GEERTS, S., '*Ambrosia maritima* L. : effet molluscicide comparé de différentes souches d'origine sénégalaise', in *Annales de parasitologie humaine et comparée,* vol. 61, n° 6, 1986, pp. 643-650.

BONNIER, Gaston, *Flore complète illustrée en couleurs de France, Suisse et Belgique,* Paris, Librairie générale de l'enseignement, 1934.

BROSSE, Jacques, *Larousse des arbres et des arbustes,* Paris, Larousse, 2000.

CARDON, Dominique, *Le Monde des teintures naturelles,* Paris, Belin, 2003.

CAZIN, François Joseph, *Traité pratique et raisonné des plantes médicinales et indigènes,* Paris, P. Asselin, 1868.

COSTES, Hippolyte (Abbé), *Flore descriptive et illustrée de la France,* Paris, Librairie des sciences et des arts, 1937.

DALÉCHAMPS, Jacques, *Histoire générale des plantes 'Sortie latine de la Bibliothèque, 1585, puis faite Françoise par Iean des Moulins, 1615. Medecins tres-fameux de leur siecle',* Lyon, Les Héritiers Guillaume Rouille, 1615.

DANTON, Philippe and BAFFRAY, Michel, *Inventaire des plantes protégées de France,* Paris, Nathan, 1995.

DELANOY, Gérard, see www.melocactus.net

DÉTHIOLLAZ, Sylvie, see www.expasy.org

D'HENNEZEL-WHITECHURCH, Marie, *Plantes médicinales d'hier et d'aujourd'hui,* Paris, Hoëbeke, 2007.

DODART, Denis, *Mémoires pour servir à l'histoire des plantes,* Paris, Imprimerie royale, 1676.

Flora Europaea, T. G. Tutin, V. H. Heywood, et al (ed.), Cambridge, Cambridge University Press, 5 vol., 1976.

Flore de Chine, 1995, see www.floras.org

FOURNIER, Paul (Abbé), *Les Quatre Flores de France,* Paris, Paul Lechevalier, 1946.

FOURNIER, Paul (Abbé), *Le Livre des plantes médicinales et vénéneuses de France,* Paris, Paul Lechevalier, 3 vol., 1948.

GIRRE, Loïc, *Les Plantes et leurs médicaments,* Paris, Delachaux & Niestlé, 2001.

HEYWOOD, V. H., *Flowering Plants of the World,* Oxford, Oxford University Press, 1978.

JOIGNEAUX, Pierre, *Le Jardin potager,* Paris, Librairie agricole de la Maison rustique, 1880

KERGUÉLEN, Michel, *Index synonymique de la flore de France,* Paris, Muséum national d'Histoire naturelle, 1993.

LE FLOC'H, Edouard, *Contribution à une étude ethnobotanique de la Tunisie,* Tunis, Imprimerie officielle de la République tunisienne, 1983.

LINNAEUS, Carolus, *Species plantarum,* 1753.

Livre rouge de la flore menacée de France, Paris, Muséum national d'Histoire naturelle et ministère de l'Environnement, vol. I (Louis Olivier, Jean-Paul Galland and Hervé Maurin, ed.), 1995.

LOBEL, Mathias (aka Lobelius) and PENA, Pierre, *Stirpium adversaria nova,* Londres, 1570.

MULLER, Serge (under the direction of), *Plantes invasives en France,* Paris, Publications scientifiques du Muséum national d'Histoire naturelle, coll. 'Patrimoines naturels', 2004.

OZENDA, Paul, *Flore du Sahara septentrional et central,* Paris, Centre national de la recherche scientifique, 1958.

PARIS, R. R.and MOYSE, H., *Précis de matière médicinale,* Paris, Masson, 30 vol., 1981.

PIGNATTI, Sandro, *Flora d'Italia,* Bologne, Edagricole, 3 vol., 1982.

POITEAU, Antoine and VILMORIN, Louis, *Le Bon Jardinier pour l'année 1844,* Paris, Audot, 1844.

RENAUX, Alain, *Le Jardin des plantes,* Florac, Parc national des Cévennes, 1988.

RENAUX, Alain, *Le Savoir en herbe. Autrefois, la plante et l'enfant,* Montpellier, Les Presses du Languedoc, 1998.

REY, Alain (under the direction of), *Dictionnaire historique de la langue française,* Paris, Le Robert, 3 vol., 1999.

ROI, Jacques, *Traité des plantes médicinales chinoises,* Paris, Paul Lechevalier, 1955.

TOURNEFORT, Joseph Pitton de, *Institutiones rei herbariae,* Joanne Anisson, 3 vol., 1700.

VILLARS, Col de, *Formules de médicaments usités dans les différens hôpitaux de la ville de Paris,* Paris, Libraire Despilly, 1768.

Mandragora officinarum L. (detail).
See p. 86

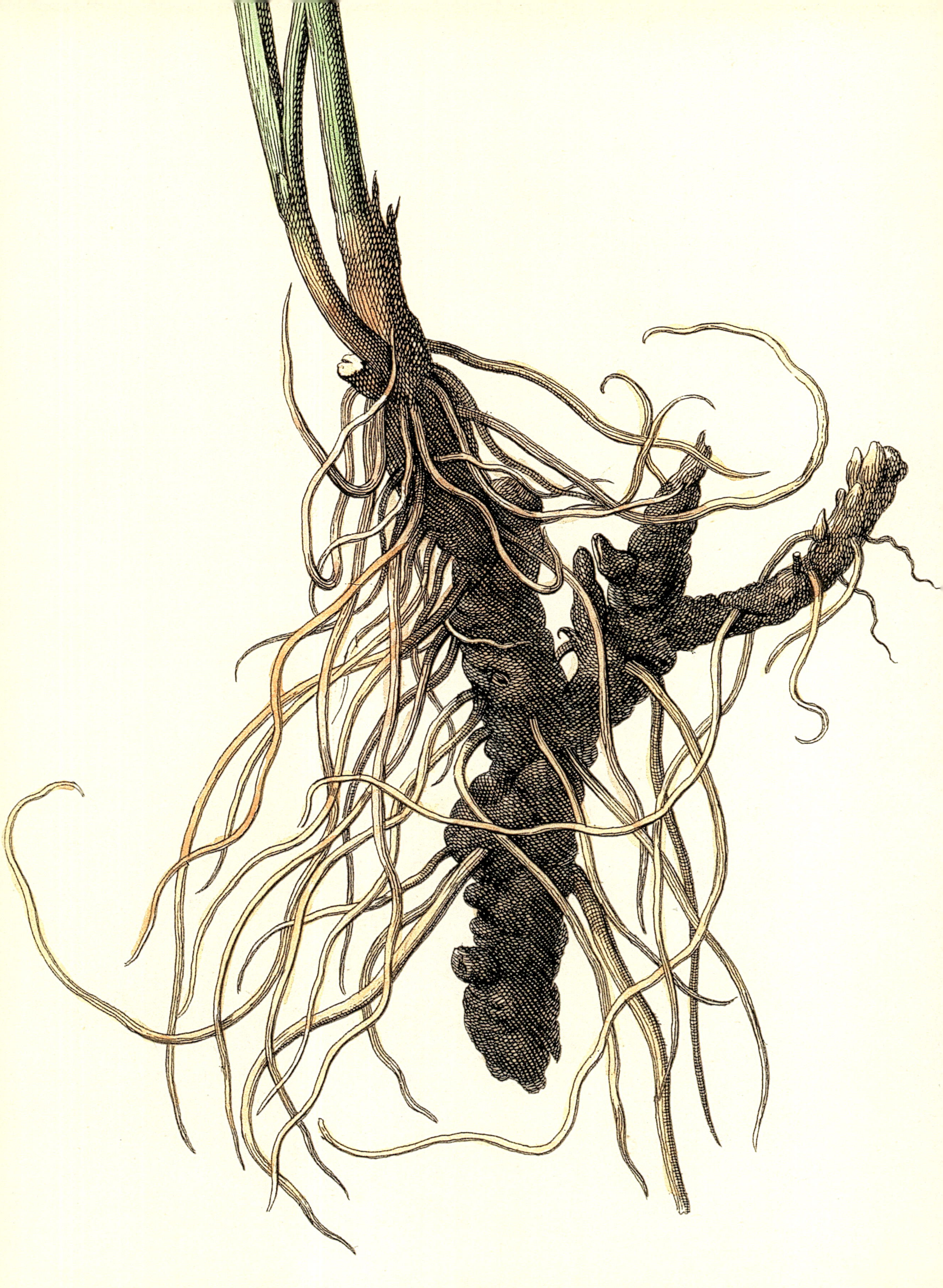

Contents

Solidago sempervirens L. (detail).
See p. 118

Photographic Credits

Réunion des musées nationaux (RMN), Louvre Chalcography Department, Paris (modern prints): pp 27, 37, 50-51, 55, 67, 81, 91, 95, 101, 103, 125, 127, 131, 137, 139.
Photos – Y. Tribes: pp 5, 11, 13, 16, 31, 35, 53, 57, 61, 87, 93, 111, 133, 140.
Photos – G. Blot: pp 39, 47, 49, 59, 63, 89, 105, 113.

Central Library, French National Museum of Natural History, Paris: pp 6, 15, 19, 22, 25, 29, 33, 41, 43, 45, 65, 69, 71, 73, 75, 77, 79, 83, 85, 97, 99, 107, 109, 115, 117, 119, 121, 123, 129, 135, 142.